Johann Caspar Rüegg (Ed.)

Peptides as Probes
in Muscle Research

With 54 Figures

Springer-Verlag

Berlin Heidelberg New York
London Paris Tokyo
Hong Kong Barcelona
Budapest

Professor Dr. Johann Caspar Rüegg
University of Heidelberg
Department of Physiology II
Im Neuenheimer Feld 326
6900 Heidelberg
FRG

QP
321
. P45
1991

ISBN 3-540-53653-1 Springer-Verlag Berlin Heidelberg New York
ISBN 0-387-53653-1 Springer-Verlag New York Berlin Heidelberg

© Springer-Verlag Berlin Heidelberg 1991
Printed in Germany

Printing: Beltz, Hemsbach; Binding: Schäffer, Grünstadt
31/3145-543210 – Printed on acid-free paper

Preface

The use of synthetic peptides as molecular probes is a new experimental approach in muscle research. "New Approaches in Muscle Physiology" was also the theme of a symposium of the German Physiological Society which was organized by R. Fink, E. Kuhn, R. Goody, and myself in Heidelberg, September 14 and 15, 1990 and supported by the Deutsche Forschungsgemeinschaft. In addition to peptides, methods using photolysis of "caged compounds" were also discussed at the symposium, but since this topic had been amply reviewed recently, it was not included in this volume. Rather, we felt that it was important to bring together, for the first time, the various peptide approaches to an understanding of muscle contraction and its regulation as well as muscle disease.

This book is dedicated to the memory of my teachers Kenneth Bailey and Hans-Hermann Weber. More than three decades ago, Weber pointed out that actin-myosin interaction is the basic event in muscle contraction, and that specific "interaction inhibitors" might be a promising tool to study this fundamental process. However, muscle contraction not only involves the interaction of the two motile proteins, actin and myosin, but in addition also a set of interacting thin filament regulatory proteins. These are troponin-C, troponin-I, troponin-T and tropomyosin. This book then describes how synthetic peptides may be used as probes to explore the cognitive interfaces between interacting muscle proteins as well as their functional significance in cardiac, smooth and skeletal muscle.

Heidelberg, Spring 1991   J.C. Rüegg

Contents

Author Index

Introduction

Johann Caspar Rüegg

Interactions between actin, myosin and regulatory proteins are pivotal in muscle activation. These interactions may now be probed by 'peptide competition' using synthetic peptides that resemble parts of the interacting protein domains. Undoubtely, the use of peptides as molecular probes is a general approach which may be used to study many kinds of protein interactions, including, for instance, those of cell adhesion proteins, or interactions of G-proteins and immunoproteins. However, muscle tissue is particularly useful to explore the power of this approach, since the effects of altering protein interactions seem to be so obvious and so readily detectable even by mechanical measurements, and, as pointed out by Schulz and Schirmer (1978), in no other tissue are the molecular details of protein structure so readily related to the overall performance of the organ!

The use of peptides to study protein function really dates back to the early studies on ribonuclease in the Rockefeller Institute in the 1960s. At that time, the idea emerged that peptide fragments of a protein, e.g. of an enzyme, may contain at least some of the biological activitiy of the whole protein molecule. Clearly, this is a "reductionist approach". However, as pointed out by Trayer (1991), "one should not assume a priori that the isolated peptides would adopt the same structure in solution as in the parent molecule, but rather that peptide fragments contain enough structural information to recognize a protein template to which they are presented", and with which they are interacting. They would then adopt the proper structure, fitting exactly into the template mould. Obviously, the mould would induce the fitting structure (induced-fit hypothesis) at the moment of interaction. At this time, NMR methods can be used to determine the structure of peptides in solution and in the bound state, and thus the requirements of the molecular recognition processes can be explored by peptide engineering. Interestingly, even small peptides containing only 10 to 20 residues may exhibit characteristic folds in free solution. In the case of transducin peptides binding to rhodopsin, there is even evidence that the free solution structure is similar to the structure of the corresponding segment in the native polypeptide chain (Hamm). In conclusion, therefore, one may be justified in using the peptide approach, hoping that protein fragments may mimic, at least partly, the biological activity of the protein as a whole. Thus, activation of myosin ATPase by actin may to some extent be mimicked by a peptide corresponding to the N-terminal end of actin (Van Eyk). Furthermore, using a peptide-NMR approach, Perry and colleagues showed that the capacity of muscle dystrophin to bind to actin resided in the extreme N-terminal region of the molecule.

In addition to the peptide mimetic approach just described, peptides may also be used as competitive inhibitors of protein interaction. Though this approach is recent in the field of muscle research, the path has been opened nearly two decades ago by immunologists, notably Atassi (1975), by showing that the interaction between antigen and antibody may be inhibited by synthetic peptides resembling the antigenic site. These peptides would compete, as it were, for the antigen recognition site on the antibody. Thus, by probing the effects of peptides corresponding to different parts of the antigen, Atassi and his colleagues were able to delineate completely the antigenic sites: Only in the case when the amino acid sequence of the peptide corresponded to that of the antigenic site, and only then, was there an inhibition of antibody-antigen interaction by peptide competition. In this way, peptides could be used as molecular probes of recognition sites involved in antigen-antibody interaction.

In a similar vein, Morita and colleagues used a heptapeptide comprising the sequence of the putative actin recognition site of myosin on the 20-kDa domain to probe the actin-myosin interface and, indeed, the peptide inhibited the formation of "rigor bonds" between actin and myosin. Recently, it also became clear that there is not just one docking site of myosin for actin. On the contrary, there are multiple parts of the myosin molecule, exhibiting an interface with actin, just as expected from A.F. Huxley's model of muscle contraction. Huxley (1980) proposed that the myosin-crossbridge surface facing the thin filament might contain many neighbouring sites interacting with actin, and that these sites would be engaged one by one during crossbridge action, thereby producing a kind of rotational movement on the actin filament. Though, at the time, this was no more than a working hypothesis, it seems clear now that crossbridges can exist in at least two conformations with low and high affinity for actin thereby forming weakly or strongly bound crossbridge states. According to Eisenberg and Hill (1985), force may be generated when crossbridges change conformation from the weak to the strong binding state. In this volume, evidence will be presented by Brenner and his colleagues that the sites of interaction between actin and myosin may be different in the weak and strong binding state. Perhaps, they comprise binding sites at the 50- and 20-kDa domain of the myosin head, respectively. In Brenner's studies, synthetic peptide probes derived from the 20-kDa or 50-kDa domain were infused into permeabilized or skinned muscle fibres in order to determine whether they affected the mechanical fibre properties that are characteristic for either the strongly or weakly bound crossbridge state. The latter is also affected by caldesmon peptides (Chalovich). Clearly as the peptide concentrations required are often high, the problem of specificity must of course always be addressed (Chase).

The idea of multi-docking sites of actin and myosin emerged from many other studies as well, including, for instance, the actin-overlay experiments of Léger and colleagues, using recombinant myosin heavy chain fragments from the 20-kDa or 50-kDa domain. It is clear, now, that the polypeptide chain around the SH-1 and SH-2 group of the 20-kDa domain as well as the C-terminal part of the 50-kDa domain and the A1-light chain all interact with actin (Trayer). Interestingly, the 20-kDa domain itself may interact with at least two functionally distinct sites of actin. By doing so, it produces either inhibition or calcium sensitization of the contractile proteins in skinned fibres as also discussed by Trayer. As through the studies of Holmes and Kabsch, the three-dimensional structure of actin is now known in detail, a steric model of myosin crossbridges interacting with actin may be within reach. Eventually, it may even be possible to describe precisely the overall movement of a muscle in terms of the molecular movements of the interacting myosin and actin molecules!

In addition to contractile proteins, interactions of regulatory proteins also play a role in muscle activation. Thus, calcium occupancy of troponin-C causes an increased interaction between troponin-I and troponin-C, whereas that of troponin-I and actin is weakened. We showed that a peptide mimic of the troponin-C recognition site of troponin-I may inhibit, competitively perhaps, the troponin-C-troponin-I interaction thereby causing relaxation of skinned muscle fibres. These effects were no longer seen after certain "point mutations" of the peptide, i.e. after replacement of certain residues by glycine.

Whereas in skeletal and cardiac muscle calcium activation involves the interaction of troponin-I with troponin-C, in smooth muscle the interaction of calmodulin with myosin light chain kinase is more important; it may be inhibited by a peptide mimic of the myosin light chain kinase amino acid sequence involved in the recognition of calmodulin (Paul and DeLanerolle). Similarly, receptor-G protein coupling may also be inhibited by peptide competition. For instance, as shown by Hamm, a peptide sequence corresponding to the rhodopsin recognition site of the G-protein transducin will competitively interfere with interaction of the proteins. Peptides interfering with G-protein interactions may also be of interest in muscle research. This is because these proteins may well play a role in regulating smooth muscle contraction and, in this respect, it may also be tempting to develop bioorganic pseudo-peptides or related drugs which mimic the action of the inhibitory peptides.

Of course, protein interactions play a role not only in signal transmission between, but also within protein molecules. For instance, the enzyme myosin light chain kinase contains a so-called pseudo-substrate sequence, which resembles the natural substrate, the phosphorylatable myosin light chain. Interestingly, the pseudo-substrate sequence competes with the substrate for the catalytic site, thereby inhibiting it. However, this inhibition is relieved, when the myosin light chain kinase binds to the calcium-calmodulin complex. As pseudo-substrate sequences have been found in many enzymes (Hardie 1988), it seems clear that peptide competition may be a natural control mechanism. Thus, the question arises whether peptides and peptide competition mechanisms may also play a physiological role in regulating intracellular signal transduction, in particular after the recent discoveries of intracellular "informational" polypeptides. It even appears that peptide hormones, which were thought to have an exclusively extracellular function, such as angiotensin, may have an effect on intracellular targets as well. Recently, Hermann-Frank discovered a peptide which is released in skeletal muscle on caffeine stimulation, but seems to have an intracellular modulatory action on the calcium release channel. How many more intracellular peptide messengers, one wonders.

In conclusion, peptides may be used much like antibodies, as molecular probes for exploring structure-function relationships of proteins. In this respect, it is well to remember that protein structures often exert a function by interacting with proteins or other macromolecules. Peptides, therefore, may either mimic the functional role of certain protein structures or, conversely, inhibit their function by peptide competition of protein interactions. In principle, peptides mimicking cognitive protein interfaces should exhibit an absolute specificity that by 'engineering' it should be possible to enhance or diminish. This offers the prospect of very selective intervention in cellular mechanism that may form the basis for a new generation of pharmacological agents.

To set the stage, it may be appropriate to begin this volume with a description of the structure of actin, which is perhaps the best-known "contractile protein" at this time. Then we proceed with contributions on the effect of peptides on contractile and regulatory proteins including peptides from G-proteins, and eventually, the final chapters of this volume will be devoted to peptide approaches to an understanding of muscle disease.

REFERENCES

Atassi MZ (1975) Antigenic structure of myoglobin: the complete immunochemical anatomy of a protein and conclusions relating to antigenic structure of proteins. Immunochemistry 12:423-438
Eisenberg E, Hill TL (1985) Muscle contraction and free energy transduction in biological systems. Science 227:999-1006
Hardie G (1988) Pseudosubstrates turn off protein kinases. Nature (London) 335:592-593
Huxley AF (1980) Reflections on muscle. The Sherrington Lectures XIV. University Press, Liverpool
Schulz GE, Schirmer RH (1978) Principles of protein structure. Springer Verlag, Berlin Heidelberg New York
Trayer J (1991) Caged peptides in Heidelberg? J Muscle Res and Cell Mot 12:216-217

Structure of Actin

Wolfgang Kabsch
Abteilung Biophysik
Max-Planck-Institut für medizinische Forschung
Jahnstraße 29
6900 Heidelberg
Germany

Almost 50 years ago actin was identified [Straub, 1942] in an extract of muscle tissue where —together with myosin— it forms the contractile protein-complex, actomyosin. Straub recognized that actin appears in two forms: In the absence of salt it can exist as a globular protein, G-actin, whereas on addition of traces of salt it changes into a highly asymmetrical fibrous protein, F-actin. Since then actin has been intensively studied in many laboratories. It was found that actin is present in all eukaryotic cells as a major component of the cytoskeleton which defines cell shape and position of internal organelles.

G-actin consists of a single polypeptide chain, a nucleotide (ATP or ADP) and a divalent cation bound to a specific site. A major achievment at the beginning seventies was the determination of the complete amino-acid sequence of actin from rabbit skeletal muscle [Elzinga et al., 1973]. Among the 375 residues are 5 free sulfhydryl residues, at positions 10, 217, 257, 285 and 374, of which Cys 374 is the most exposed residue. It is a favourite target for labelling with fluorescent dyes and for fluorescence resonance energy transfer measurements. Acetylation of the N-terminal aspartate and methylation of histidine 73 are posttranslational modifications. Since then actin from many other sources has been sequenced. It has become apparent that actin is a highly conserved molecule throughout eukaryotic evolution. The largest differences are found in the sequence of actin from the ciliate *Tetrahymena* which is still 75% homologous with rabbit skeletal muscle actin [Hirono et al., 1987].

Actin forms a tight 1:1 complex with bovine pancreatic deoxyribonuclease I (DNase I) and thereby looses its capability to polymerize in the presence of salt [Lazarides and Lindberg, 1974]. The biological significance of this complex is still obscure but it was found to be useful for growing crystals suitable for X-ray diffraction analysis. The 3-dimensional structure of the actin:DNase I complex has been determined

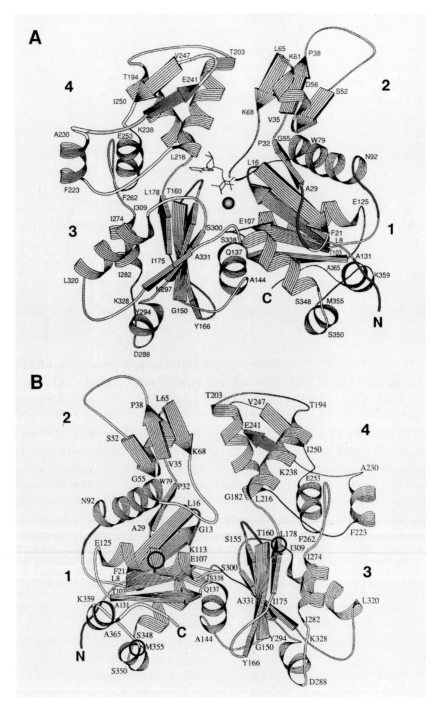

Fig. 1

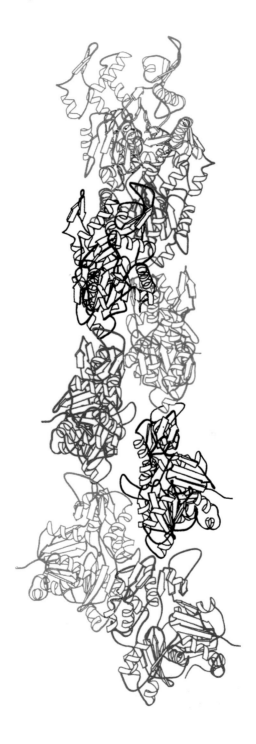

Fig. 2

recently [Kabsch et al., 1990] at 2.8Å resolution providing detailed information about the spatial position of all atoms and the functional regions of the molecule. Moreover, it was possible to derive the orientation of the actin monomers in the F-actin filament [Holmes et al., 1990]. In this article we give a short description of the actin structure expecting this to be useful as a reference frame for some of the contributions to this symposium.

A schematic picture of actin as found in the complex with DNase I is shown in Fig. 1. First and last aminoacid residues of helices and β-pleated sheet strands are specified. Actin is a flat molecule of width 35Å that fits into a square of sidelength 55Å. It consists of two domains (for historical reasons called large and small, although they turn out to be of almost the same size) with the nucleotide (ATP or ADP) and an associated calcium ion bound in the cleft between the two domains. Both N- and C-termini are in the small domain. Each of the domains consists of two subdomains. Subdomains 1 (residues 1–32, 70–144, and 338–372) and 2 (residues 33–69) constitute the small domain while subdomains 3 (residues 145–180, and 270–337) and 4 (residues 181–269) represent the large domain of actin. The motif of a 5-stranded β-sheet consisting of a β-meander and a right handed $\beta\alpha\beta$-unit occurs in each domain (subdomains 1 and 3). 75 pairs of corresponding residues have been identified which can be superimposed by an approximate twofold rotation with a root-mean-square error of 2.8Å. This suggests that gene duplication might have occured although there is very little similarity between the equivalent amino acid residues.

Fig.1: Two views of the schematic representation of the structure of actin. The molecule consists of 2 domains which are assembled from two subdomains each. The small domain is built from subdomains 1 and 2. The large domain consists of subdomains 3 and 4. ATP and the high-affinity Ca^{2+} ion are found in the cleft between the domains as shown in Fig. 1A. First and last aminoacid residues of helices and β-pleated sheet strands are specified. Functional regions in the molecule are shown in color: yellow regions are involved in actin-actin contacts in the filament; regions in subdomain 1 which are known to be involved in binding of myosin subfragment 1 (S1) are shown in red; the two nucleotide phosphate binding loops are in blue; regions which are likely to be involved in binding tropomyosin are painted green. Fig. 1B shows the back side of actin in the same orientation as in 1A. Binding sites for phalloidin at Met 355, Glu 117, Met 119 (Vandekerckhove et al., 1985) and residue Arg 177 which can be ADP-ribosylated by clostridial toxins (Aktories and Wegner, 1989) are indicated by circles.

Fig. 2: Schematic representation of the F-actin helix. Eight monomers are shown. For clarity adjacent molecules are in different colors. The barbed end of the filament is at the bottom in the picture. Subdomains 1 and 2 of actin are at large radius.

The major interaction between actin and DNase I involves residues 39–45, 61 and 63 in subdomain 2. Residues Gly 42, Val 43, and Met 44 of actin are incorporated as a parallel strand of a β-sheet in DNase I with partners Tyr 65, Val 66, and Val 67. The minor contact involves residues Glu 207 and Thr 203 in subdomain 4 which interact with His 44 and Glu 13 of DNase I. The tight binding causes a distortion of the DNase I structure [Oefner and Suck, 1986] from its native form without actin by as much as 1.8Å in the C_α-postions of residues 65–67. It is likely then that the actin structure is also disturbed in the regions of subdomain 2 which are in close contact with DNase I. X-ray analysis of crystals of the actin:profilin complex is in progress [Schutt et al., 1989], and this is expected to provide an undistorted structure of subdomain 2 since the contact region between these two proteins is near Glu 364 in subdomain 1 [Vandekerckhove et al., 1989].

ADP or ATP are located at a strategic position in the cleft between the domains thereby contributing to the stability of the structure. The β- and γ-phosphate oxygens form hydrogen bonds with amides of the β-bends in subdomain 1 (residues Ser 14, Gly 15, and Leu 16) and subdomain 3 (residues Asp 157, Gly 158, Val 159), respectively. The ribose is in the 2'-endo conformation. The 2'- and 3'-hydroxyl groups of the ribose participate in hydrogen bonds with one of the oxygen atoms of the carboxylate groups of Glu 214 and Asp 157, respectively. The adenine sits in a pocket formed by residues Lys 213, Glu 214, Thr 303, Met 305, Tyr 306, and Lys 336. The high affinity calcium ion is found in a hydrophilic environment built from Gln 137, Asp 11 and 154. The ion is bound to the nucleotide β- and γ-phosphates.

F-actin can be orientated in capillary tubes and the orientated gels yield characteristic X-ray fibre diffraction patterns with a resolution of up to 8Å [Popp et al., 1987]. The atomic model of the ADP-actin monomer was placed in the F-actin helix in all possible orientations and its computed X-ray fibre diffraction pattern was compared with the measured fibre diagram. A unique orientation of the monomer was found which fits the measured data very well [Holmes et al., 1990]. The diameter of the filament is about 90Å. As shown in fig. 2 the large domain of actin is at small radius and the small domain is at large radius. The interactions between molecules along the two-start actin helix involve residues 322–325 with 243–245; 286–289 with 202–204; 166–169 and 375 with 41–45. Contacts along the left-handed genetic helix are provided between residues 110–112 and 195–197. The only other strong contacts across the axis possibly involves the loop 262–272 which contains a number of exposed hydrophobic residues in the actin crystal structure. Rebuilding this loop into an antiparallel β-sheet leads to a finger-like structure inserting across the helix axis into the hydrophobic pocket formed by residues 166, 169, 171, 173, 285, 289, 63, 64, and residues 40–45. The distance between the C_α-positions of Lys 191 and Cys 374 on adjacent monomers along the left-handed genetic helix is 19Å in the atomic model which agrees with the estimate obtained by cross-linking different actin monomers in the filament [Elzinga and Phelan, 1984].

The structures of F-actin and of actin decorated with myosin S1 as obtained by cryoelectron microscopy have been reported recently [Milligan et al., 1990] at a resolution of 25Å. If the atomic model of the F-actin filament is reduced to the same resolution the agreement between the two structures is very good. Moreover, Milligan et al have labelled Cys 374 with undeca-gold and obtained a distance of 27Å from the filament axis for this label; in the atomic model it is at 25Å which is also consistent with fluorescence energy transfer measurements. The elongated S1-molecule binds to actin in a tangential manner resulting in the characteristic arrow-head appearence which has been used for the definition of filament polarity. As revealed from the cryoelectron microscopic images the major contact region is on the outer (small) domain of actin which is well in line with the identification of known myosin binding residues at positions 1–4, 24–28, 93, 95 in the actin sequence. In addition, these images show that the N-terminal portion of the myosin light chain 1 binds to actin at a location containing residues 360–363. In the presence of Ca^{2+} and S1 the images show that tropomyosin follows the long pitch helix and lies at a filament radius of 38Å. Comparison with the atomic model of the filament suggests that tropomyosin runs in the groove between subdomain 3 and 4 with contacts around Lys 215 and Pro 307. The helix Asp 222 – Ser 233 appears also to be part of this contact. This is in agreement with the observation that the sequence of *Tetrahymena* actin [Hirono et al., 1987], which does not bind muscle tropomyosin [Hirono et al., 1990], differs in 8 amino acids in the region 222–233 from the rabbit skeletal muscle sequence.

The atomic structure of actin has yielded a synthesis of results obtained from biochemical and immunological studies as well as from biophysical methods such as NMR, fluorescence energy transfer, electronmicroscopy, fibre diffraction, and X-ray crystallography. Moreover, structural analysis of actin has revealed some rather surprising similarities with other important proteins in the cell: hexokinase [Steitz et al., 1976] and the ATPase domain of the heat shock protein HSC70 [Flaherty et al., 1990]. Both proteins have the same topology as found in subdomains 1 and 3 of actin. In the case of the HSC70 ATPase fragment, whose structure has been solved recently [Flaherty et al., 1990], the structural similarity includes in addition subdomains 2 and 4 of actin. 241 pairs of equivalent aminoacid residues have been identified in the two structures which can be superimposed with a root-mean-square distance of 2.3Å [Flaherty et al., 1991]. Furthermore, the conformation of the adenine nucleotides and their environments are very similar. The structural differences between the two molecules mainly occur in loop regions of actin known to be involved in interactions with other monomers in the actin filament or in binding myosin or tropomyosin. We believe that the structural similarity between actin and HSC70 is not accidental: it suggests the existence of a common ancestral molecule along the evolutionary pathway which should also include hexokinase. The branching into actin and the heat shock proteins must have occured a very long time ago since there is little sequence similarity between the two: Only 39 out of 241 structural equivalent aminoacid residues are identical. The strongest sequence homology consists of 7 matches out of 24 residues in the actin region Ala 295–Thr 318 which is part of the adenine binding pocket. In addition, there is a sequence "fingerprint" of the nucleotide β-phosphate binding loop which is present in actin, bovine HSC70, yeast hexokinase B, rat liver glucokinase, and E. coli glycerol kinase.

Acknowledgement: I would like to thank Drs. Goody, Holmes, and Pai for helpful comments and suggestions in preparing this manuscript.

References

Aktories, K. and Wegner, A.
J. Cell Biol. **109**, 1385–1387 (1989).

Elzinga, M., Collins, J.H., Kuehl, W.M. and Adelstein, R.S.
Proc.Natl.Acad.Sci.U.S.A. **70**, 2687–2691 (1973).

Elzinga, M. and Phelan, J.J.
Proc.Natl.Acad.Sci. U.S.A. **81**, 6599–6602 (1984).

Flaherty, K.M., DeLuca-Flaherty, C., McKay, D.
Nature **347**, 623–628 (1990).

Flaherty, K.M., McKay, D., Kabsch, W. and Holmes, K.C.
Proc.Natl.Acad,Sci. U.S.A.; in press (1991).

Hirono, M., Endoh, H., Okada, N., Numata, O. and Watanabe, Y.
J Mol Biol, **194**, 181–192 (1987).

Hirono, M., Tanaka, R. and Watanabe, Y.
J Biochem **107**, 32–36 (1990).

Holmes, K.C., Popp, D. , Gebhard, W. and Kabsch, W.,
Nature **347**, 44–49 (1990).

Kabsch, W., Mannherz, H.G., Suck, D., Pai, E.F. and Holmes, K.C.
Nature **347**, 37–44 (1990).

Lazarides, E. and Lindberg, U.
Proc.Natl.Acad.Sci.U.S.A. **71**, 4742–4746 (1974).

Milligan, R.A., Whittaker, M. and Safer, D.
Nature **348**, 217–221 (1990).

Oefner, C. and Suck, D. J.Mol.Biol. **192**, 605–632 (1986).

Popp, D., Lednev, V.V., Jahn, W. J. Mol. Biol. **197**, 679–684 (1987).

Schutt, C.E., Lindberg, U., Myslik, J. and Strauss, N.
J.Mol.Biol. **209**, 735–746 (1989).

Steitz, T.A., Fletterick, R.J., Anderson, W.A., and Anderson, C.M.
J.Mol.Biol. **104**, 197–222 (1976).

Straub, F.B. Studies, University of Szeged **II**, 3–15 (1942).

Vandekerckhove, J.S., Deboben, A., Nassal, M. and Wieland, T.
EMBO J. **4**, 2815–2818 (1985).

Vandekerckhove, J.S., Kaiser, D.A. and Pollard, T.D.
The Journal of Cell Biology **109**, 619–626 (1989).

Interaction of Actin 1–28 with Myosin and Troponin I and the Importance of These Interactions to Muscle Regulation

J.E. Van Eyk, F.D. Sönnichsen, B.D. Sykes, and R.S. Hodges

Department of Biochemistry, University of Alberta, Edmonton, Alberta, Canada T6G 2H7

This investigation was supported by research grants from the Medical Research Council of Canada and the Alberta Heart Foundation and by an Alberta Heritage Foundation for Medical Research Studentship (J.V.E.) and Fellowship (F.D.S.).

INTRODUCTION

Muscle contraction involves the regulation of the interaction between actin and myosin (acto-myosin) by the regulatory protein complex, troponin-tropomyosin (Tn-TM). The troponin (Tn) subunits (TnC, TnI, TnT) interact with each other and work in concert with actin-TM to produce conformational changes which alter the acto-myosin interaction. When F-actin binds the myosin head, activation of the ATP hydrolysis rate occurs causing an increase in the release of Pi, thereby increasing the rate of cross-bridge cycling.

Several synthetic peptides which mimic biologically important regions of various muscle proteins have been successfully used to study the mechanism of muscle contraction. The studies by Talbot and Hodges (1979 and 1981a,b), which initiated the extensive synthetic peptide research on muscle proteins, showed that the synthetic peptide of TnI, TnI 104-115, is the minimum sequence required to inhibit the ATPase activity. The TnI peptide also mimics TnI by possessing TM specificity and the ability to bind TnC which results in the release of inhibition (Cachia *et al.*, 1983, 1986; Van Eyk and Hodges, 1987). Eleven analogs of the TnI peptide in which a single glycine residue was substituted at each position of the sequence showed that all amino acids, but especially the C-terminal residues, were essential for full biological activity (Van Eyk and Hodges, 1988). As well, the TnI peptide competes with TnI and inhibits force development in skinned muscle fibres (Rüegg *et al.*, 1989). Recently, the 3D structure of TnI 104-115 bound to TnC in the presence of Ca^{2+} has been determined by NMR (Campbell and Sykes, 1990a).

Synthetic peptides have also been used to study the role of TnC in muscle regulation. The interaction of metal ions with a wide variety of synthetic peptide analogs of a Ca^{2+} binding site of TnC (loop or helix-loop-helix) were studied in an attempt to understand the variation in metal specificity and affinity and the effect of different residues on the conformation of the loop region. This includes a series of analogs of the loop region containing all possible combinations of Asp and Asn in the first three Ca^{2+} coordinating ligand positions. It was concluded from studies with these analogs that metal affinity is a balance of attractive forces between ligand and metal-ion and the repulsive forces between the negatively charged ligands so that the relative spatial distribution of the charges are important (Gariépy et al., 1982, 1983, 1985; Reid et al. 1980, 1981; Marsden et al., 1988, 1989). As well, a 34-residue peptide representing a single Ca^{2+} binding unit (loop-helix-loop) forms a symmetric two-site dimer upon Ca^{2+} binding which is similar in tertiary structure to the C-terminal domain of TnC (Shaw et al., 1990).

Synthetic peptides having the amino acid sequence around the reactive thiol (SH1) of myosin (702-708) bind F-actin (Suzuki et al., 1987; Eto et al., 1990) and completely inhibit the acto-S1 ATPase activity (Suzuki et al., 1990). Work using larger synthetic myosin peptides from the region encompassing SH1, residues 690-725, indicate there were two functionally distinct actin binding sites contained within this region (Keane et al., 1990).

Finally, the synthetic peptide approach has been extended to the interactions of actin with other muscle proteins. Immunological data, crosslinking, and NMR experiments have strongly indicated that the N-terminal region of actin contains one of the binding sites between actin and myosin (Sutoh, 1982, 1983, Moir and Levine, 1986; Moir et al. 1987; Méjean et al. 1986, 1987; Miller et al., 1987). As well, the N-terminal region of actin has been implicated in the acto-TnI interaction, in particular TnI fragment 96-116 (Levine et al., 1988, Grabarek and Gergely, 1987). In this paper synthetic peptides of the N-terminal region of skeletal actin were used to probe the acto-myosin and acto-TnI interactions. These actin peptides are an excellent example of how the structure and function of a peptide and the analogous region in the native protein are related.

RESULTS

Interaction Between Actin 1-28 and Myosin Fragments

A peptide of skeletal actin, actin 1-28, and three analogs were synthesized using standard procedures for solid-phase peptide synthesis (Table 1). Actin 1-28 activated the myosin subfragment 1 (S1) ATPase activity to 150% (Fig. 1, Panel A, Van Eyk and Hodges, 1990), indicating that this peptide binds S1 and alters the S1 conformation such that the ATP hydrolysis rate is increased. However, low levels of precipitation occurred under these assay conditions (Fig. 1, Panel B). The precipitate was composed of equivalent quantities of S1 and peptide suggesting that a 1:1 complex is formed which is less soluble than S1.

TABLE 1.   Amino acid sequences of the N-terminus of skeletal actin and the
            synthetic actin peptides

Skeletal Actin

```
                       10                    20              28
                        *
       Ac-D E D E T T A L V C D N G S G L V K A G F A G D D A P R . . .
```

Actin Peptides

1-28

```
                             *
       Ac-D E D E T T A L V A D N G S G L V K A G F A G D D A P R-amide
```

1-14

```
                             *
       Ac-D E D E T T A L V A D N G S-amide
```

8-20

```
                           *
             Ac-L V A D N G S G L V K A G-amide
```

18-28

```
                Ac-K A G F A G D D A P R-amide
```

---

*   The cysteine at residue 10 in the native actin sequence was substituted by alanine in the synthetic peptides.

Actin 1-28 did not precipitate the larger myosin fragment, heavy meromyosin (HMM) or S1 when present as the acto-S1 complex (1:1 mole ratio) (Van Eyk and Hodges, 1990). Actin 1-28 bound S1 and HMM and activated the ATPase activities in a manner similar to F-actin, at mole ratios below 1 equivalent of peptide to S1, or 0.5 equivalent of peptide to HMM (Fig. 2, panel A and B, respectively). Analysis of this data assuming Michaelis kinetics showed that the K_{app} (0.4 X 10^6 M) and V_{max} (162-173%) were identical for actin 1-28 interaction with S1 and HMM. At higher ratios of F-actin or actin 1-28 to S1 or HMM, only F-actin continued to activate the ATPase activities in a concentration dependent manner. The peptide reached its maximum activity at a ratio of 0.5:1 peptide to HMM, indicating that at least one HMM head had peptide bound. This is

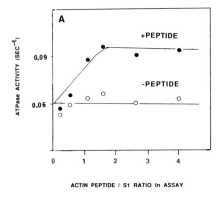
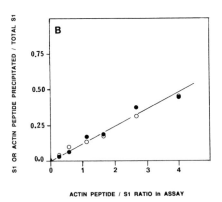

Figure 1. Interaction between actin peptide 1-28 and S1. The S1-ATPase activity in the absence (O) and presence (●) of the actin 1-28 is shown in panel A. The concentration of S1 was 2 μM. ATPase assays were performed using an automatic pH stat apparatus (Van Eyk and Hodges, 1988). A 2 ml assay sample in an Mg^{2+}-ATPase buffer consisting of 2 mM Tris, 0.1 mM EGTA, 5 mM $MgCl_2$, 30 mM KCl and 2.5 mM ATP, pH 7.6, was used. Centrifugation of the assay sample was carried out at 15,000 rpm for 15 mins. The quantitation of S1 (O) and peptide (●) in the precipitate determined by analysis of the supernatent (S1, amino acid analysis; peptide, radioactivity) is shown in panel B.

not surprising since a significant fraction of HMM is bound to F-actin with only one head in the presence of Mg^{2+}-ADP (Duong and Reisler, 1987) suggesting that cooperative interactions occur between the two heads. By comparison, S1, which consists of a single head and binds only one actin monomer (Sutoh, 1983; Chen et al. 1985; Greene 1984; Heaphy and Tregear, 1984), reached its maximum activity at approximately 1:1 ratio of actin peptide to S1. Therefore, formation of the F-actin filament is required for further activation of the ATPase to occur. Numerous studies have suggested that the binding of S1 and presumably HMM to F-actin filament induces long range conformational changes along the filament (Ikkai, et al., 1979, Yanagida et al., 1984 and Rouagrene et al., 1985). Therefore, the increased ATPase activity after 1:1 binding of actin to S1 or HMM may reflect the cooperativity within the actin filament.

In order to determine the minimum sequence required for full biological activity and the relative importance of the regions within actin 1-28, several truncated analogs (1-14, 8-20 and 18-28) were synthesized (Table 1). Actin 1-28 was the most effective and potent activator suggesting actin residues 1-28 is the minimum sequence required for maximum biological activity (Fig. 2, panels C and D, Van Eyk and Hodges, 1991). Two non-overlapping analogs, actin 1-14 and actin 18-28, which represent the N- and C-terminal regions of actin 1-28, showed high activities which were similar to each other in both the acto-S1 and HMM ATPase assays. Actin 8-20 which covers the central section was considerably less potent. These results suggest that the N- and C-terminal regions of actin 1-28 are equally important for the binding of actin to myosin and the activation of the ATPase activity.

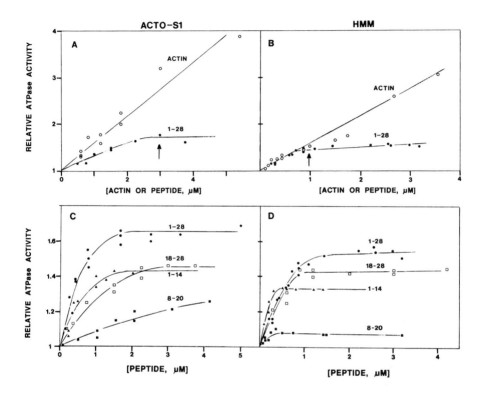

Figure 2. The effect of F-actin and actin peptides on the acto-S1 (panel A and C) and HMM (panel B and D) ATPase activities. The concentration of F-actin and S1 was 1:1 and each protein was 3 µm in panel A or 2.5 µm in panel C. The acto-S1 ATPase activity in the absence of actin peptide(s) or additional F-actin was taken to be 1.0. Acto-S1 ATPase assays were performed under identical conditions as the S1 ATPase assays in figure 1. The concentration of HMM was 2.0 µm in panel B and 2.5 µm in panel D. The HMM activity in the absence of actin peptide(s) or F-actin was taken to be 1.0. HMM ATPase assays were performed under identical conditions as the S1 ATPase assays in figure 1 except, the assay buffer consisted of 2 mM Tris, 30 mM KCl, 2 mM MgCl₂, 0.1 mM EGTA, 2.5 mM ATP, pH 8.0.

Interaction Between Actin 1-28 and Tn

In order to determine whether actin 1-28 could interact with Tn, high performance size-exclusion chromatography (SEC) was performed. The actin peptide 1-28 bound Tn in the absence but not the presence of Ca^{2+}(Fig. 3). The actin 1-28/troponin complex (peak 1) was easily separated from free peptide (peak 2) by size-exclusion chromatography. Analyses of peak 1 by reversed-phase microbore chromatography (panel C), where all components were separated, clearly shows that actin peptide is eluted with troponin due to complex formation. In the presence of Ca^{2+}, troponin did not bind actin 1-28 as shown in the analysis of peak 3 by reversed-phase chromatography (panel D). The peptide was eluted in peak 4 (compare panel D and F). Further, SEC demonstrated that the actin peptide was bound to the TnI subunit of the Tn complex (data not

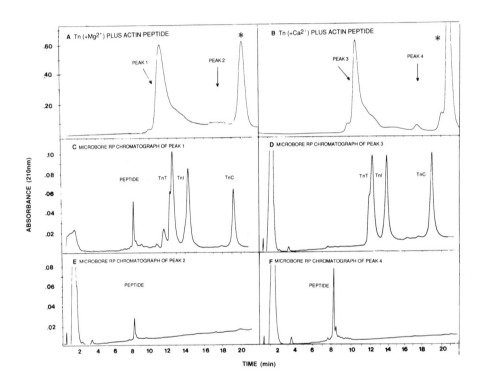

Figure 3. Actin peptide 1-28 and Tn interaction determined by size-exclusion chromatography. Panel A and panel B, respectively shows the elution profile of a mixture of actin peptide and Tn in the absence of Ca^{2+} (presence of Mg^{2+}) and presence of Ca^{2+}, at a 1:1 mole ratio (3 nmole of each component). The peptide-protein mixtures were run on a high performance size-exclusion column, Altex TSK G2000 SW (7.5 mm I.D. x 30 cm) in a buffer consisting of 10 mM Tris, 100 mM KCl, pH 6.8 at a flow rate of 0.5 ml/min. A peak (*) corresponds to the BME that was present in the Tn sample to ensure the cysteine residues of TnI and TnC remained in the reduced state, was eluted following the free peptide (panel A and B). The various peaks (1 to 4) were collected and analyzed on a reversed phase microbore column (Brownlee, Aquapore RP-300, 7µ, 1.0 mm I.D. x 100). A linear AB gradient (2% B/min) (where solvent A is 0.05% aqueous TFA and solvent B is 0.05% TFA in acetonitrile at a flow rate of 0.1 ml/min) was used to separate the 3 subunits of Tn from the actin peptide (panel C to F).

Table 2. Retention of Actin Analogs during TnI affinity chromatography[a]

Actin peptide	Elution conditions[b] (mM KCl)
1-28	176
1-14	160
8-20	no binding
18-28	85

[a] Analogs were chromatographed on a Beckman Ultra-affinity EP column (4.6 mm I.D. x 5 cm) derivatized with skeletal TnI. Peptides were eluted by applying a linear AB gradient of 1.5% B/min which is equivalent to 15 mM/min, where solvent A was 20 mM Tris, 0.1 mM EGTA, pH 6.8 and solvent B was solvent A plus 1.0 M KCl. The flow-rate was 0.5 ml/min and the effluent was monitored at 210 nm.

[b] KCl concentration required to elute a peptide was calculated from the retention time minus t_g (gradient delay time) times the gradient rate of 15 mM KCl/min, for example with actin 1-28 the retention time was 17.5 min and t_g was 5.8 min = (17.5 - 5.8) x 15 = 176 mM.

shown). This indicates that interaction between actin 1-28 and TnI is sensitive to the Ca^{2+}-dependent changes in TnC. High performance affinity chromatography employing a skeletal TnI derivatized column was used to determine the strength of the interaction between the various actin analogs and TnI. The actin analogs (1-14, 8-20 and 18-28) bound less tightly to the TnI affinity column than actin 1-28, suggesting that residues 1-28 of actin are essential for TnI interaction (Table 2). Actin 1-14 bound considerably tighter to TnI than actin 18-28 while actin 8-20 did not bind at all. Unlike the S1 and HMM interaction, the N-terminal region of 1-28 is more important than the C-terminal region for the interaction between actin and TnI.

Previous work which indicated that the N-terminal region of actin binds to residues 96-116 of TnI (Levine et al., 1988), was confirmed since the synthetic TnI peptide, TnI 104-115, could successfully compete with TnI (derivatized to the HPAC column) for the actin 1-28. TnI 104-115 is the minimum sequence required to inhibit the TM-acto-S1 ATPase activity (Talbot and Hodges, 1981; Van Eyk and Hodges, 1988). This peptide also binds to TnC. Another TnC-binding peptide, mastoparan (a basic tetradecapeptide from wasp venom) was able to compete actin 1-28 for the TnI affinity column but was considerably less effective than the TnI peptide. Both the TnI peptide and mastoparan are proposed to bind to helix E of TnC (N-terminal helix of Ca^{2+} binding site III) (for a review, see Cachia, et al., 1985).

The Structure of Actin 1-28

Having shown that the peptide binds to troponin and myosin (S1 and HMM) and is also biologically active, we were interested in determining the structure of the peptide in solution and when bound to these proteins. In the recently published X-ray crystallographic structure obtained from G-actin/DNAase I crystals (Kabsch et al., 1990) the first 8 N-terminal residues and a short sequence between F^{21} and A^{29} are shown to be on the surface and relatively unstructured. Therefore these residues are accessible and could be directly involved in the acto-myosin or acto-troponin interactions. Between residues L^8 and F^{21} a β-sheet was indicated leading from the surface into the core of the protein and, after a beta-turn between G^{13} and L^{16}, back to the surface again. These two strands are stabilized by additional strands on either side to form a five stranded, β-pleated sheet.

The Solution Structure of the Free Peptide

1D- and 2D-NMR spectroscopy was employed to determine the solution structure of the peptide. The proton resonances were assigned using standard procedures of spin system identification (DQF-COSY, ^{1}H-^{1}H RELAY, TOCSY) and sequential connectivity determination (NOESY, ROESY). Analyzing the chemical shifts, it is striking that almost all values are close to those for a random coil (Bundi and Wüthrich, 1979). Thus, 12 out of 28 CαH-protons resonate in the range of 4.33 ± 0.05 ppm. Only a few residues show significant deviation from random coil values. These are G^{23} and A^{26} and to a smaller extent L^8, V^9, L^{16} and V^{17}. Furthermore the CαH resonances of G^{13} and G^{23} are non-degenerate. Further structural evidence was obtained from studies of the temperature dependance of the chemical shifts. Some amide resonances show lowered shifts upon a change in temperature, which indicates hydrogen bonding. The lowest observed shift was 3 ppb/°C for G^{23} while several other shifts were lower than 6 ppb/°C, which is indicative of structure in the regions D^3-T^5, G^{13}-L^{16} and G^{23}-A^{26}.

More detailed information was obtained through 2D-NOESY spectra; a list of connectivities is given in Fig. 4, Section A. The spectra contain strong sequential and medium to small intraresidue $d_{N\alpha}$ crosspeaks, but hardly any medium range crosspeaks were detected. A strong crosspeak between A^{26}Hα -P^{27}Hδ characterizes the trans-proline peptide bond. A small number of d_{NN}-connectivities support the existence of structure in the three regions mentioned above, but since the number of crosspeaks is fairly small the amount of information obtained is not sufficient to further characterize these regions of the peptide.

The remaining parts of the peptide, (T^5-A^7, L^8-G^{13}, V^{17}-A^{22}) are flexible and in an extended conformation. The NOESY-spectrum lacks any $d_{\alpha\alpha}$ - or side chain contacts. These contacts would be expected to appear for a β-sheet comparable to the one formed in the crystal structure. Furthermore, some observed sequential sidechains contacts (L^8V^9 and $L^{16}V^{17}$) are incompatible with a β-sheet, where sidechains of adjacent residues are placed on opposite sides of the sheet. Additionally, no alternately lowered amide exchange rates and no increase in $^3J_{NH}$ can be observed. Consequently, though we can't exclude the formation of a beta-turn between G^{13} and L^{16}, the double stranded β-sheet structure is not present in this peptide in solution. This is not

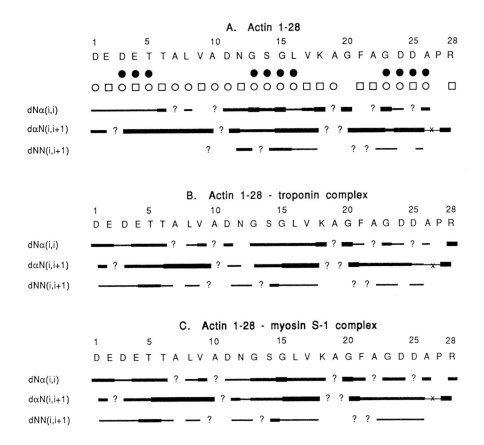

Figure 4. Summary of sequential connectivities, coupling constants and temperature shift data. The intensity of the nOes is indicated by the thickness of the lines, grouped into strong, medium and weak according to distances < 2.7, < 3.3 < 5 Å. Sequential connectivities involving $P^{27}\delta H$ are indicated by an **X**, while overlapping crosspeaks are indicated by a ?. Resonances marked with a filled circle show a temperature shift $\Delta\delta < 6$ ppb/°C. An open circle represents a coupling constant $^3J_{NH\alpha} < 7$ Hz, and an open square $7 < {}^3J_{NH\alpha} < 9$ Hz.

surprising, since in G-actin the antiparallel double strand of residues L^8 to F^{21} is stabilized through further strands on each side, a stabilization which is missing in the case of the shortened peptide.

The Bound Structure of Actin 1-28

To obtain structural information about the peptide when bound either to troponin (in absence of Ca^{2+}) or myosin, transferred NOESY (trNOESY) experiments were performed. This experiment is the expansion of the 2D-NOESY technique to exchanging systems such as ligand-protein complexes. Assuming fast exchange and a low protein concentration (5 to 10 % relative to peptide concentration) chemical shifts and linewidth are dominated by the concentration of the free peptide (p^f).

$$p^f \gg p^b \qquad \delta_{obs} = p^f \delta^f + p^b \delta^b$$
$$\Delta\upsilon_{obs} = p^f \Delta\upsilon^f + p^b \Delta\upsilon^b$$

Conversely, the crosspeak intensity in trNOESY spectra is dominated by the bound peptide, since it depends not only on the concentration but also on the correlation times of free or bound peptide. Using small peptides the correlation time of the free species is close to the extreme narrowing limit with a cross relaxation rate close to zero. The bound peptide adopts the correlation time of the protein, leading to large cross relaxation rates.

$$\omega_{ij}{}^b \gg \omega_{ij}{}^f$$
$$a_{ij}(\tau_m) = - [p^b \omega_{ij}{}^b + p^f \omega_{ij}{}^f] \tau_m \approx - [p^b \omega_{ij}{}^b] \tau_m$$

Thus, strong negative nOe's conveying structural information of the bound peptide are transferred to free ligand via fast exchange and detected through the intense, narrow lines of the free form. Fig. 5 demonstrates the effect of the chemical exchange on the efficiency of the magnetization transfer, i.e., the crosspeak intensity. Comparing the crosspeak intensity of actin 1-28 in solution and in trNOESY experiments, the effective cross relaxation time is dramatically shortened leading to more intense crosspeaks. The relative change of crosspeak intensity and the detection of new connectivities are due to altered distances in the bound form of the peptide and therefore an effect of the interaction.

The connectivities listed in Fig. 4 (sections B and C) show minimum change in crosspeak intensity in the region between G^{13} and G^{20}, thus reflecting the same conformation in all three spectra, i.e. free in solution and bound to either troponin or S1. This region is either not involved in binding or not affected by it. The largest change occur in the N-terminal part of the peptide, in particular between D^1 and A^7, suggesting that this is the major site of interaction. Together with a reduction of sequential $d_{\alpha N}$-, and an increase in intraresidue $d_{N\alpha}$- crosspeaks a row of d_{NN}-

connectivities were detected. This result suggests the formation of a helix in this part of the peptide, but since no medium range (dαN(i,i+3)) connectivities were observed, we interpret these results as a transition from an extended structure towards a helix such as a series of turns or a loosened, distorted helix. Interestingly this region (1-7) shows identical nOe-patterns when the peptide is bound to either troponin or S1.

Differences between the free and bound form, as well as between the two bound structures were detected in the regions D^{11}-N^{12} and F^{21}-R^{28}. The latter region forms a second binding site, which is structured differently as indicated by the nOe intensities for D^{24}-D^{25} and R^{28}. Since the binding of the N-terminal part of the peptide to both proteins is identical, the conformational differences in this region (F^{21}-R^{28}) may reflect the difference in the observed binding affinities of the peptide towards troponin and S1 on a structural level.

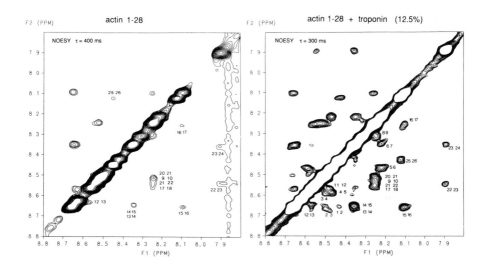

Figure 5. Region of amide-amide connectivities of actin 1-28 and actin 1-28/troponin.

DISCUSSION

One of the fundamental questions in biochemistry is the relationship between structure and function of active sites of proteins. In recent years, peptides have been successfully used to mimic the activities of proteins, being excellent probes to study protein-protein interactions of complex multi-protein systems. Furthermore, combining bioassays with modern NMR structural methods is a successful approach to correlate structure and function.

Actin 1-28 showed a high activity in the bioassays and proved to be one of the binding sites between actin and troponin or myosin. It consists of three distinct regions. The central region, as shown in the assays using actin 8-20, has a minimal effect on the binding and the biological activity (Fig. 2, Table 2). This is consistent with the finding that the conformation of the central residues 8-20 (with one exception) is not affected by binding to either troponin or S1. This indicates that the central region does not bind or has only minimal contact with the proteins. The N-terminal region of actin 1-28, represented by actin peptide 1-14, strongly interacts with troponin and myosin. This peptide binds tightly to a TnI affinity column and is critical for the activation of the S1 and HMM ATPase activity. This strong interaction induces a conformational change, especially in the first 7 residues of actin 1-28. Interestingly, the conformation is identical when bound to either troponin or S1, which implies that the actin-binding site of TnI and S1 for the N-terminal residues of actin are similar.

The C-terminal region of actin 1-28 (actin 18-28) binds to both proteins as well. In the case of actin binding S1 or HMM, the C-terminal interaction is as important as the binding of the N-terminus, while it is less important when actin 1-28 binds to troponin. This difference is observed on a structural level. While binding of the peptide to either protein induces a change in the conformation of residues 21-28, thus proving the direct involvement of these residues in binding, the conformation of this region in the complex with either protein is not the same. This reflects the differences of the actin binding sites of TnI and myosin and the ability of actin 1-28 to adapt and fit these two proteins.

Another small difference between the two bound conformations was detected at residues D^{11}-N^{12}. The significance of this change is so far unclear. Since the change was limited to these two residues, it seems unlikely that this change is due to peptide-protein interaction in this region. We assume that the conformational change is caused by the binding of the protein to the N-terminal binding site of the peptide and reflects small differences in the protein conformation close to the peptide-protein interface. This might be an 'experimental artefact' of using a synthetic peptide, which shows unusual flexibility in the central region of the peptide. In the intact protein these two residues are buried in the core of the protein as a part of a five stranded β-sheet and are therefore unlikely to change conformation. Another interpretation would be that D^{11}-N^{12} serves as a hinge region meeting different distance requirements between the N-terminal and the C-terminal binding sites of the peptide when bound either to TnI or S1. For the intact protein this would imply, that binding to troponin or S1 not only leads to a conformational change at the N and C-terminal binding sites, but also alters the central β-sheet structure. This would affect the conformation of the whole subdomain.

The finding that actin 1-28 can interact with troponin I and myosin fragments has implications for understanding muscle contraction. The N-terminal region of actin 1-28 acts as a chemical switch between two sites of attachment, showing identical structure upon interaction with Tn and S1. As shown by Levine *et al.* (1988) and in this study (by the competitive release of actin 1-28 from the TnI affinity column) the N-terminus of actin interacts with the peptide comprising residues 104-115 of TnI. This region of TnI is also a chemical switch in muscle regulation showing Ca^{2+}-dependent interaction between actin ($-Ca^{2+}$) or TnC ($+Ca^{2+}$). One can now speculate that the TnI binding site of actin and TnC show similarities. This is probable considering that the residues of the TnI peptide which are important for binding of the peptide to TnC are also critical for inhibition of the S1 ATPase activity (Van Eyk and Hodges, 1988). The TnI peptide binds to the C-domain of TnC, in particular helix E, which is the N-terminal helix of Ca^{2+} binding site III (see Cachia *et al.*, 1985 for review, Leszyk *et al.*, 1990; Campbell *et al.*, 1989; Campbell and Sykes, 1990b). Upon binding of the TnI peptide, there is an increase in α-helical content of TnC and a pCa shift such that lower Ca^{2+} concentrations are required to fill the regulatory low affinity Ca^{2+} binding sites (Rüegg *et al.*, 1989, unpublished results). Similarily, the binding of TnI to actin 1-28 induces a conformational change from an extended structure to a helix-like conformation between residues 1-7, the region which is profoundly affected by the binding of Tn. Since the TnI peptide contains 6 positive charges and the amino terminus of actin 4 negative charges, coulombic interactions seem to play a major role in this interface. By analogy the actin

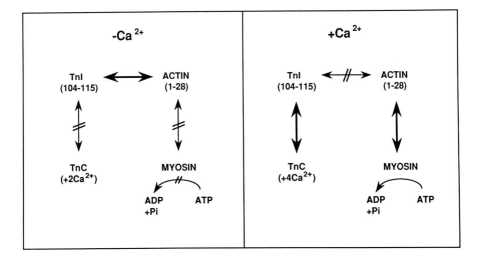

Figure 6. Protein interactions during muscle regulation.

(1-7) binding site of S1 should reflect the binding site of TnI, that is, it should be positively charged and mimic the 3D-structure of the TnI inhibitory region (104-115) (Campbell and Sykes, 1990a and b).

The regions of actin, residues 1-28, and TnI, residues 104-115, are two of the Ca^{2+}-dependent switches that control muscle relaxation and contraction (Fig. 6). The TnI inhibitory peptide, TnI 104-115, which interacts with TM-actin causing inhibition of S1 ATPase activity, also binds to TnC. As well, the actin peptide 1-28 which interacts with S1 (or HMM) causing activation of the S1 ATPase activity, also binds TnI, in particular residues 104-115. In the absence of Ca^{2+} when either Mg^{2+} or Ca^{2+} are bound to the two high affinity Ca^{2+} binding sites of TnC, the interaction between TnI and actin (plus TM) is dominant. Muscle relaxation is then promoted due to inhibition of the ATPase activity (Van Eyk and Hodges, 1986). In contrast, when Ca^{2+} binds to the two regulatory low affinity Ca^{2+} binding sites on TnC, TnI inhibition is released since TnI 104-115 has a stronger binding affinity for TnC in the presence than in the absence of Ca^{2+} (Van Eyk and Hodges, 1988) and actin 1-28 cannot bind Tn in the presence of Ca^{2+}. Actin 1-28 then interacts with S1 causing potentiation of the ATPase activity and muscle contraction. The interactions described in Fig. 6 do not take into account the role of TM. The presence of TM in the ATPase assay greatly enhances the activity of TnI and TnI 104-115 (Talbot and Hodges, 1979, Eaton *et al.*, 1975) through multi-site interactions with up to 7 actin monomers in filamentous actin.

In summary, cyclic association-dissociation of actin and myosin coupled to ATP hydrolysis is the most essential process in muscle contraction. The Tn complex regulates the acto-myosin interaction by altering which proteins interact with one another depending whether Ca^{2+} is absent or present. Work using synthetic peptides of actin and TnI have clearly demonstrated that actin residues 1-28 and TnI residues 104-115 are involved in the chemical switches which controls the interactions between TnC, TnI, actin and myosin.

REFERENCES

Bundi A, Wüthrich K (1979) ^{1}H-NMR Parameters of the Common Amino Acid Residues Measured in Aqueous Solutions of the Linear Tetrapeptides H-Gly-Gly-X-Ala-OH, Biopolymers 18:285 - 297

Cachia PJ, Gariépy J, Hodges RS (1985) in Calmodulin Antagonist and Cellular Physiology; Structural studies on calmodulin and troponin C: Phenothazine, peptide and protein interactions with Ca^{2+}-induced helices (Hidaka, H, and Hartshorne, NJ, eds), pg 63-88, Academic Press, NY

Cachia PJ, Sykes BD, Hodges RS (1983) Calcium-dependent inhibitory region of troponin: a proton nuclear magnetic resonance study on the interaction between troponin C and the synthetic peptide N^{α}-Acetyl[FPhe[106]]TnI-(104-115) amide. Biochemistry 22:4145-4152

Cachia PJ, Van Eyk J, Ingraham RH, McCubbin WD, Kay CM, Hodges RS (1986) Calmodulin and troponin C: a comparative study of the interaction of mastoparan and troponin I inhibitory peptide [104-115]. Biochemistry 25:3553-3562

Campbell AP, Marsden BJ, Van Eyk J, Hodges RS, Sykes BD (1990) in Frontiers of NMR in Molecular Biology. Calcium regulation of muscle contraction; NMR studies of the interaction of troponin C with calcium and troponin I (Live, D, Armitage, I, Patel, D, Vol 109, ed) pg 99-108, Alan R. Liss, Inc.

Campbell AP, Sykes BD (1990a) Interaction of troponin I and troponin C: use of the two-dimensional transferred nuclear overhauser effect to determine the structure of the inhibitory troponin I peptide when bound to turkey skeletal troponin C. J Mol Biol submitted

Campbell AP, Sykes BD (1990b) Interaction of troponin I and troponin C: ^{19}F NMR studies of the binding of the inhibitory troponin I peptide to turkey skeletal troponin C. J Biol Chem submitted

Chen T, Applegate D, Reisler E (1985) Cross-linking of actin to myosin subfragment 1: Course of reaction and stoichiometry of products. Biochemistry 24:137-144

Duong AM, Reisler E (1987) The binding of myosin heads on heavy meromyosin and assembled myosin to actin in the presence of nucleotides. J Biol Chem 262:4129-4133

Eaton BL, Kominz DR, Eisenberg E (1975) Correlation between the inhibition of the Acto-heavy meromyosin ATPase and the binding of tropomyosin to F-actin: Effects of Mg^{2+}, KCl, troponin I and troponin C. Biochemistry 14:2718-2725.

Eto M, Suzuki R, Morita F, Kuwayama H, Nishi N, Tokura S (1990) Roles of the amino acid side chains in the actin-binding S-site of myosin heavy chain. J Biochem 108:499-504

Gariépy J, Kay LE, Kuntz ID, Sykes BD, Hodges RS (1985) Nuclear magnetic resonance determination of metal-proton distances in a synthetic calcium binding site of rabbit skeletal troponin C. Biochemistry 24:544-550

Gariépy J, Sykes BD, Hodges RS (1983) Lanthanide-induced peptide folding: variations in lanthanide affinity and induced peptide conformation. Biochemistry 22:1765-1772

Gariépy J, Sykes BD, Reid RE, Hodges RS (1982) Proton nuclear mganetic resonance investigation of synthetic calcium-binding peptides. Biochemistry 21:1506-1512

Grabarek Z, Gergely J (1987) Location of the TnI binding site in the primary structure of actin. Acta Biochim. Biophys Hung 22 (2-3) 307-316

Green LE (1984) Stoichiometry of actin S1 crosslinked complex. J Biol Chem 259:7363-7366

Heaphey S, Tregar R (1984) Stoichiometry of covalent actin-subfragment 1 complexes formed on reaction with a zero-length crosslinking compound. Biochemistry 23:2211-2214

Ikkai T, Wahl P, Auchet J (1979) Anisotropy decay of labelled actin, evidence of the flexibility of the peptide chain in F-actin molecules. Eur. J. Biochem. 93:397-408

Kabsch W, Mannherz HG, Suck D, Pai EF, Holmes KC (1990) Atomic structure of Actin : DNAase 1 complex. Nature 347:37-44

Keane AM, Trayer IP, Levine BA, Zeugner C, Rüegg JC (1990) Peptide mimetics of an actin binding site on myosin spanning two functional domains on actin. Nature 344:265-268

Leszyk J, Grabarek Z, Gergely J, Collins JH (1990) Characterization of zero-length cross-links between rabbit skeletal muscle troponin C and troponin I: evidence for direct interaction between the inhibitory region of troponin I and NH_2-terminal, regulatory domain of troponin C. Biochemistry 29:299-304

Levine BA, Moir AJG, and Perry SV (1988) The interaction of troponin-I with the N-terminal region of actin. Eur J Biochem, 172:389-397

Marsden BJ, Hodges RS, Sykes BD (1988) ^{1}H NMR studies of synthetic peptide analogues of calcium-binding site III of rabbit skeletal troponin C: Effect on the lanthanum affinity of the interchange of aspartic acid and asparagine residues at the metal ion coordinating positions. Biochemistry 27:4198-4206

Marsden BJ, Hodges RS, Sykes BD (1989) A ^{1}H NMR determination of the solution conformation of a synthetic peptide analogue of calcium-binding site III of rabbit skeletal troponin C. Biochemistry 28:8839-8847

Mejéan C, Royer M, Labbé J-P, Derancourt J, Benjamine Y and Roustan R (1986) Antigenic probes locate the myosin subfragment 1 interaction site on the N-terminal part of actin. Bio Sci Rep 6:493-499

Mejéan C, Boyer M, Labbé JP, Marlier L, Benjamin Y, Roustan C (1987) Anti-actin antibodies an immunulogical approach to the myosin-actin and tropomyosin-actin interfaces. Biochem J 244:571-577

Miller L, Kalnoski M, Yunossi Z, Bulinski JC, Reisler E (1987) Antibodies directed against N-terminal residues on actin do not block acto-myosin binding. Biochemistry 26:6064-6070

Moir AJG, Levine BA (1986) Protein cognitive sites on the surface of actin: A proton NMR study. J Inorg Biochem 27:271-278

Moir AJG, Levine BA, Goodearl AJ, Trayer JP (1987) The interaction of actin with myosin subfragment 1 and with pPDM-cross-linked S1: a ^{1}HNMR investigation. J Muscle Res Cell Motl 8:68-69

Moir AJG, Levine BA (1983) Studies of the interaction of troponin I with proteins of the I-filament and calmodulin. Biochemical J 209:417-426

Reid RE, Clare DM, Hodges RS (1980) Synthetic analog of a high affinity calcium binding site in rabbit skeletal troponin C. J Biol Chem 255:3642-3646

Reid RE, Gariépy J, Saund AK, Hodges RS (1981) Calcium-induced protein folding. J Biol Chem 256:2742-2751

Rouayrenc J, Bertrand R, Kasab R, Walzhöny D, Bädler M and Walliman T (1985) Further characterization of the structural and functional properties of the cross-linked complex between F-actin and myosin S1. Eur J Biochem 146:391-401

Rüegg JC, Zeugner C, Van Eyk J, Kay CM, Hodges RS (1989) Inhibition of TnI-TnC interaction and contraction of skinned muscle fibres by the synthetic peptide TnI [104-115]. Pflügers Arch 414:430-436

Shaw G, Hodges RS, Sykes B (1990) Calcium-induced peptide association to form an intact protein domain: ^{1}H-NMR Stuctural Evidence. Science 249:280-283.

Sutoh K (1982) Identification of myosin-binding sites on the actin sequence. Biochemistry 21:3654-3661

Sutoh K (1983) Mapping of actin-binding sites on the heavy chain of myosin. Biochemistry 22:1579-1585

Suzuki R, Nishi N, Tokura S, Morita F (1987) F-actin-binding synthetic heptapeptide having the amino acid sequence around the SH1 cysteinyl residue of myosin. J Biol Chem 262:11410-11412

Suzuki R, Morita F, Nishi N, Tokura S (1990) Inhibition of actomyosin subfragment 1 ATPase activity by analog peptides of the actin-binding site around the Cys(SH1) of myosin heavy chain. J Biol Chem 265:4939-4943

Talbot JA, Hodges RS (1979) Synthesis and biological activity of an icosapeptide analog of the actomyosin ATPase inhibitory region of troponin I. J Biol Chem 254:3720-3723

Talbot JA, Hodges RS (1981b) Comparative studies on the inhibitory region of selected species of troponin-I. J Biol Chem 256:12374-12378

Talbot JA, Hodges RS (1981a) Synthetic studies on the inhibitory region of rabbit skeletal troponin I. J Biol Chem 256:2798-2802

Van Eyk JE, Cachia PJ, Ingraham RH, Hodges RS (1986) Studies on the regulatory complex of rabbit skeletal muscle: Contributions of troponin subunits and tropomyosin in the presence and absence of Mg^{2+} to the Acto-S1 ATPase activity. J Prot Chem 5:335-354

Van Eyk JE, Hodges RS (1987) Calmodulin and troponin C: affinity chromatographic study of divalent cation requirements for troponin I inhibitory peptide (residues 104-115), mastoparan, and fluphenazine binding. Biochem Cell Biol 65:982-988

Van Eyk JE, Hodges RS (1988) The biological importance of each amino acid residue of the troponin I inhibitory sequence 104-115 in the interaction with troponin C and tropomyosin-actin. J Biol Chem 263:1726-1732

Van Eyk JE, Hodges RS (1990) A synthetic peptide of the N-terminus of actin interacts with myosin, submitted

Van Eyk JE, Hodges RS (1991) Synthetic peptide studies on the interaction of the N-terminal region of actin with myosin subfragment 1 (S1) and heavy meromyosin (HMM) submitted

Yanagida T., Nakase M. Nishigama K, and Oosawa F (1984) Direct observation of motion of single F-actin filaments. Nature (London) 307:58-60

Probing Myosin Head Structure with Monoclonal Antibodies and Recombinant Technology

P.Eldin, J.O.C.Léger, M.LeCunff, B.Cornillon, M.Anoal, D.Mornet, H.P.Vosberg[*] and J.J.Léger

Institut National de la Santé et de la Recherche Médicale, INSERM U300, Faculté de Pharmacie, Av.Ch.Flahaut, 34060 Montpellier
[*]Max Planck Institue for Medical Research, Department of Cell Biology, Jahnstrasse 29, D-6900 Heidelberg, FRG

Myosins from mammalian striated muscles are hexameric oligomers composed of two heavy chains (MHC) and two pairs of different light chains (LC1 and LC2). Differences in the primary structures of the myosin subunits and various assembly combinations may allow fine modulations of the different myosin properties, such as ATPase activities, affinity for actin, mutual interactions between the light and heavy subunits and MHC associations in thick filaments. These structural variations would thus finally modulate the contractile properties of the corresponding muscles. Physiopathological effectors, like excessive hormone impregnation, mechanical overloading,... could induce dramatic changes in the myosin subunit composition and alter the contractile properties of the affected muscles (Swynghedauw, 1986).

Cardiac isomyosins currently provide a very convenient and simple model for investigating the structure-function relationships between different sets of heavy and light myosin subunits. Indeed cardiac isomyosins are known to be composed of only 2 different main heavy chains (α and β MHC), and of only 2 different pairs of light chains, one specific to the ventricles and the other to the atria. 5 or 6 different subunit assemblages are found in cardiac muscles :

Ventricles:		Atria:	
$\alpha\alpha$		$\alpha\alpha$	
	VLC1		ALC1
$\beta\beta$ +		$\beta\beta$ +	
	VLC2		ALC2
$\alpha\beta$		$(\alpha\beta)$?	

Figure 1: Molecular Hexameric Organisation of cardiac isomyosins. α and β MHCs are associated with VLC1 and VLC2 in ventricles, and with ALC1 and ALC2 in the atria. They form homodimers in both cardiac chambers, but the heterodimeric association has only been detected in the ventricles (Dechesne et al., 1987).

It has been shown that the calcium-dependent ATPase activity of $\alpha\alpha$ isomyosins is 3-fold higher than that of $\beta\beta$ isomyosins, and that the actin affinities of both isomyosins differ in proportionally similar way. Ventricles containing only αα isomyosins have significantly higher contraction velocity than ventricles containing only ββ isomyosins. In fact, the spatial distribution of these cardiac isomyosins varies within the ventricles and the atria, inducing subtle alterations in the local contractile properties (for review, see Swynghedauw et al.,1990). The amino acid sequences of a few α and β MHCs have recently been derived from the corresponding isolated cDNAs. The 2 isomyosins from rat heart are very similar (92% at the nucleic level and 93 % at the amino acid level). The differences were unexpectedly distributed in discrete areas along the molecule. Some of the difference clusters are located within or near postulated functional sites such as the ATP or actin binding sites (Mornet et al., 1989), and the MLC binding sites located near the S1-S2 junctions (Winkelman et Lowey, 1986):

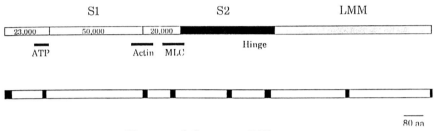

■ Clusters of Sequence Differences

Figure 2. Sequence comparison between rat cardiac α and β MHC. (for the ß human MHC, Jaenicke et al. 1990, for the rat α and β MHC, McNally et al, 1989). S1 = Subfragment 1; S2 = Subfrgament 2; LMM = Light Meromyosin

It is important to determine how such small structural differences induce the activity differences that have been observed between both cardiac isomyosins which play a very subltle role in regulating cardiac performance. We have recently developped new tools to investigate the structure-function relationship within the different cardiac isomyosin molecules. In this paper, we present two complementary approaches, one based on monoclonal antibodies specific for both heavy and light subunits and another based on genetic engineering techniques, which have been applied in the study of cardiac ß isomyosins.

1. ATPase alterations of bovine cardiac ß isomyosin by monoclonal anti-Subfragment 1 (S1) antibodies.

Monoclonal antibodies (Mab) were prepared from mice immunised with purified native bovine cardiac ß myosin S1. The Mabs, which were directed against different epitopes and classified into 5 subgroups corresponding to their positions along the S1 heavy chain, served as functional probes by interfering or not with basic myosin functions such as ATPase activities, actin binding and MLC interactions. We describe an example of this approach with two anti-S1 monoclonal antibodies, the epitopes of which were located at the 23-50K junctional region (group 4 in Fig.3). Antibodies C8 and H10 had very different effects on various myosin ATPase activities. Mab C8 had no effect on the ß isomyosin calcium-dependent ATPase but severely inhibited the magnesium-dependent actin-activated ATPase activity of the corresponding ß myosin subfragment 1. In contrast, Mab H10 severely inhibited the calcium-dependent ATPase of ß isomyosin and strongly activated the magnesium-dependent actin-activated ATPase activity of the corresponding ß myosin subfragment 1.

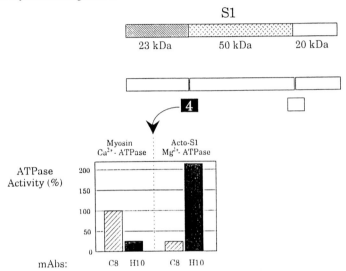

Figure 3. Alterations of some enzymatic activities of ß isomyosins by 2 monoclonal antibodies against the 25-50K junction. Both Mabs were purified from hybridoma culture supernatants using protein A MemSepCartridges (Millipore). Their respective specificities were determined by Western blotting on tryptic S1 subfragments (23K, 50K and 20K). Cardiac bovine ß isomyosin and chymotryptic S1 were incubated for 30 min. at 4°C in the presence or absence of antibody, in buffers that were used for further ATPase activity measurements. ATP hydrolysis was initiated by adding either ATP to myosin, or F-Actin to S1 for 5 min. before adding ATP. Inorganic phosphate releases were measured by the Malachite green method. Results are expressed, as percent of myosin or S1 activity in the absence of antibody. The molar ratio of monoclonal antibody/myosin or S1 is equal to about 2.

This experiment confirms that the 23-50K junctional region is highly accessible to specific Mabs and plays a crucial role in the ATP hydrolysis process. It also suggests that ATP hydrolysis differs according to the local conformation of the myosin fragment used (entire molecule or chymotryptic fragment,etc...) and to various other cofactors (ions, F-

36

actin,etc…). If the same functional Mabs were applied on cardiac α isomyosins, this would allow us to investigate the effects which different protein sequences at the 23-50K junction have on the interactions of antibodies with cardiac isomyosin.

2. Mapping of the F-actin binding site on the cardiac human ß myosin heavy chain.

E.coli cDNA expression systems were developed from different clones coding for human ß cardiac MHC. The functionality of recombinant MHC fragments were then characterized through competition assays with myosin subfragment-1 in F-actin cosedimentation and acto-S1 ATPase measurements. Direct experiments including F-actin overlay, MLC overlay and ATP binding, were also performed (Eldin et al., 1990).

We present here the end result of mapping of the F-actin binding site(s) on the cardiac human ß myosin heavy chain, obtained using fusion proteins containing different overlapping recombinant ß MHC fragments:

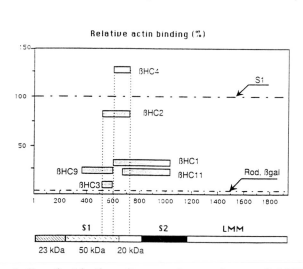

Figure 4. F-actin binding site mapping on human ß MHC recombinant fragments. Different cDNA fragments coding for cardiac ß MHC were inserted in expression plasmids pEx and gave rise in E.coli to six proteins (called ßHC1, 2, 3, 4, 9 and 11) fused with a large ß galactosidase moiety. The six fusion proteins containing different overlapping recombinant ß MHC fragments were separated on SDS-PAGE, then electrotransferred on nitrocellulose sheets and submitted to a urea-based renaturation procedure.The sheets were finally incubated in a freshly prepared solution of F-actin at 1 mg/ml. The actin-myosin interactions were revealed by the fixation of anti-actin antibodies, then identified by a alkaline phosphatase-coupled rabbit anti-mouse antibody. Actin binding intensities were measured by densitometric scanning and compared to that

observed with the homologous cardiac human ß chymotryptic S1 which was taken as 100 %. Fusion proteins are presented according to their linear extension along the human cardiac ß MHC amino acid sequence. Rod portions of myosin or ß-galactosidase were taken as negative controls for actin binding.

These results are in agreement with previous studies suggesting that the connecting region between the 50K and the 20K subregion of S1 is the main site of actin binding in MHC. The proposed approach allows an interesting dissection of the MHC regions implicated in actin binding. The interaction of the whole S1 with F-actin could result from the addition of MHC segments that positively or negatively react with F-actin. For example, ßHC2, which is composed of one negative ßHC3 element and one positive ßHC4 element, has an intermediary relative interaction with actin. It is now crucial that we obtain a spatial picture of these interactive sites. We hope to be able to detect differences between both cardiac isomyosins, using the same approach with human cardiac α isomyosin.

We are presently trying to develop new expression systems to obtain recombinant soluble MHC fragments that could be more directly applied in the near future in functional activity assays and biophysical approaches such as nuclear magnetic resonance or cristallographic analysis. In addition to the use of antibodies as phenotypical, conformational and even functional probes, we consider the recombinant technology approach as a powerful new way for investigating the molecular mechanism of isomyosins and their role in the energy transduction process during muscle contraction.

Acknowledgments: This work was supported by INSERM, CNRS, l'Association Française contre les Myopathies et la Caisse National d'Assurance Maladie des Travailleurs Salariés. P.E. was awarded a fellowship from the Ministère de la Recherche.

References:
Dechesne, C. A., Bouvagnet, P., Walzthöny, D and Léger, J. J. (1987) J. Cell. Biol. 105, 3031-3037.
Eldin, P., LeCunff, M., Diederich, K. W., Jaenicke, T., Cornillon, B., Mornet, D., Vosberg, H.P. and Leger, J. J. (1990). J. Musc. Res. Cell Motility 11, 378-391
Jaenicke, T., Diederich, K.W., Haas, W., Schleich, J., Lichter, P., and Vosberg, H.-P. (1990). Genomics 8, 194-206.
Mc Nally, E. M., Fraft, R., Bravo-Zehnder, M., Taylor, D. A., and Leinwand, L. A. (1989). J. Mol. Biol. 210, 665-671.
Mornet, D., Bonet, A., Audemard, E., and Bonicel, J. (1989). J. Musc. Res. Cell Motility 10, 10-24.
Swynghedauw, B. (1986). Physiol. Rev. 66,710-771.
Swynghedauw et al. (1990) Research in cardiac hypertrophy and failure. John Libbey & Company Ltd/INSERM. London
Winkelmann, D.A. and Lowey,S. (1986) J.Mol.Biol. 188, 595-612

An Actin-Binding Site on Myosin

Fumi Morita, Tsuyoshi Katoh, Rie Suzuki, Katsuyoshi Isonishi,
Kouichirou Hori, and Masumi Eto
Department of Chemistry
Faculty   of   Science
Hokkaido University
Sapporo, Kita-ku
Hokkaido 060
Japan

In order to understand the molecular mechanism of muscle
movement or cell motility, identification and characterization
of actin-binding sites on myosin head and myosin-binding sites
on actin molecules are necessary.   Elucidation of the
correlation between the binding sites and the myosin ATPase
reaction is also essential.   We will discuss here an actin-
binding site located around the reactive Cys(SH1)(705) on the
myosin heavy chain and its role in the actomyosin ATPase
reaction.

It is known that when rabbit skeletal muscle myosin S-1 is
incubated with a bifunctional reagent, N,N'-p-phenylene
dimaleimide (pPDM), the two reactive thiol groups on the S-1
heavy chain, Cys(SH1) and Cys(SH2), are chemically crosslinked
[Reisler et al., 1974, Burke & Knight, 1980, Wells et al.,
1980].   The crosslinking reaction was found to be inhibited
strongly in the presence of F-actin [Katoh et al., 1984].   The
results indicated that an actin-binding site may be present
between the two SH groups, and that the bound F-actin may
sterically block the crosslinking.   In order to examine this
possibility, the smallest BrCN peptide containing the two thiol
groups was isolated and renatured by dialysis against a sucrose
solution [Katoh et al., 1985].   The renatured peptide, having
a molecular weight of 10,000, bound F-actin with a dissociation
constant of about 1 μM [Katoh et al., 1985].

Next, several oligopeptides having the amino acid sequence
in the vicinity of Cys(SH1) were chemically synthesized [Suzuki
et al, 1987, Morita et al., 1987].   Dissociation constants of
these peptides with F-actin were determined from the degree of
competitive inhibition against the formation of acto-S-1 rigor

Peptide	Dissociation constant (mM)
(SH1)	
(1) Val-Leu-Glu-Gly-Ile-Arg-Ile-Cys-Arg-Lys-Gly-OEt	0.19 ± 0.07
(2) Ile-Arg-Ile-Cys-Arg-Lys-Gly-OEt	0.23
(3) Val-Leu-Glu-Gly-Ile-Arg-Ile-OMe	0.87
(4) Ile-Arg-Ile-OMe	1.1
(5) Cys-Arg-Lys-Gly-OEt	12
(6) Val-Leu-Glu-Gly-OEt	30

Table I   Dissociation constants of acto-peptide complexes. Conditions were 20 mM MOPS-KOH (pH 7.0), 0.1 M KCl, and 50 mM 2-mercaptoethanol at 4 °C.   1.6 to 3.0 µM F-actin and S-1 in the range of 0.3 to 2.8 µM were used.

complex.   As shown in Table I, the dissociation constants of peptides (1) and (2) had almost the same values.   The dissociation constants of peptides (3) and (4) also had values similar to each other.   The N-terminal extra tetrapeptide segment of peptides (1) and (3), Val-Leu-Glu-Gly, does not appear to contribute significantly to the binding with F-actin. Therefore, the heptapeptide, peptide (2), is critical for this binding.   In this heptapeptide, the N-terminal Ile-Arg-Ile segment appears to be more important for the binding with F-actin than the C-terminal Cys-Arg-Lys-Gly segment since the affinity for F-actin of tripeptide (4) was about 10 times higher than that of tetrapeptide (5).   Peptide (4) has one positive charge in the side chains, while peptide (5) has two charges.   Therefore, the positive charges are not determinant for the affinity.   The sequence of amino acids seems to be critical.

Various substituted heptapeptides were synthesized and their abilities to bind with F-actin were compared [Eto et al., 1990].   The hydrophobic side chains of Ile(702) and Arg(703), the positive charge of Arg(703), and the sulfur atom of Cys(705) on the S-1 heavy chain were determined to be critical, but the side chain of Ile(704) was not. Keane et al. [Keane et al., 1990] have reported, using longer synthetic peptides, that

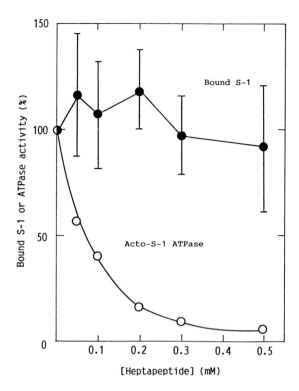

Fig. 1   Effects of the heptapeptide, Ile-Arg-Ile-Cys-Arg-Lys-Gly-OEt, on the acto-S-1 ATPase activity (O) and on the binding between S-1 and F-actin (●). Conditions were 2 mM $MgCl_2$, 2 mM ATP, 0.1 mg/ml bovine serum albumin, and 20 mM Tris-HCl (pH 8.0) at 25 °C. The ionic strength was adjusted to 10 mM by adding KCl. The binding was determined by an ultracentrifugal separation method (Suzuki et al., 1990).

the polypeptide segment closer to the C-terminal than this region is also important for the binding of S-1 with F-actin.

In order to determine the correlation between the actin binding site of this region (S-site) and the ATPase reaction, effects of the heptapeptide (peptide (2)) on S-1 ATPase activity were examined [Suzuki et al., 1990].   The heptapeptide inhibited the ATPase activity in the presence of F-actin to the level of that of S-1 alone (Fig. 1).   The S-1 ATPase activity in the absence of F-actin, however, was not affected at all [Suzuki et al., 1990], suggesting that the

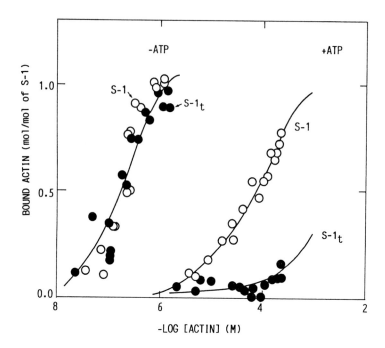

Fig. 2   Binding of F-actin to S-1 and S-1$_t$. Conditions were 10 mM NaCl, 2 mM MgCl$_2$, 0.1 mg/ml bovine serum albumin, and 20 mM Tris-HCl (pH 8.0) in the absence and presence of 2 mM ATP at 18 to 20 °C.   The open and closed circles were obtained with S-1 and S-1$_t$, respectively.   The F-actin binding with S-1 was determined by an ultracentrifugal separation method (Katoh & Morita, 1984).

heptapeptide is correlated with the activation process by F-actin, but not directly correlated with the catalytic process of ATPase.

The amount of S-1 bound to F-actin during the steady state of acto-S-1 ATPase was determined by an ultracentrifugal separation method.   As shown in Fig. 1, the amount of the bound S-1 was changed only slightly by the addition of the heptapeptide even when the acto-S-1 ATPase activity was inhibited completely.   These results suggest two important possibilities: 1) the S-site is necessary to bind with F-actin to activate the S-1 ATPase activity, and 2) another site different from the S-site maintains the acto-S-1 link during the steady state of the ATPase reaction.   This second actin-binding probably occurs in the J-site, located around the junctional region between the 50-kDa and the 20-kDa domains of the S-1 heavy chain [Yamamoto & Sekine, 1979, Mornet et al., 1979, Sutoh, 1983, Chaussepied & Morales, 1988].

The degree of F-actin-binding to S-1 and to S-1$_t$, in which the J-site was digested with trypsin, was determined by an ultracentrifugal separation method [Katoh & Morita, 1984]. Figure 2 replots data from the paper by Katoh & Morita. In the absence of ATP, both S-1 and S-1$_t$ bound F-actin with almost the same affinities. In the presence of ATP, on the other hand, the affinity of S-1 decreased more than 100 times, and that of S-1$_t$ decreased even more. These results suggest that the strong affinity of the acto-S-1 rigor complex is determined at the S-site, which is located far from the site of the tryptic digestion. In the presence of ATP, on the other hand, the affinity seems to be determined at the J-site, since the degree of actin-binding was decreased markedly by cleavage at the J-site. The strongly binding S-site, therefore, is suggested to be dissociated in the presence of ATP.

The dissociation of the S-site in the presence of ATP was further suggested from the reactivity of SH1 on S-1 to the fluorescent reagent, 5-(iodoacetamidoethyl)aminonaphthalene-1-sulfonic acid (IAEDANS). Figure 3 shows the time courses of the reactions of IAEDANS with the S-1 heavy chain. Since the molar concentration of the bound fluorescent reagent was less than that of the S-1 used, the reagent seemed mainly to bind with the reactive SH1 group. Under conditions where S-1 alone can react with IAEDANS, as shown in Fig. 3a, the acto-S-1 rigor complex could not react with the reagent at all (Fig. 3b). The acto-S-1 complex in the presence of ATP, however, reacted with the reagent to almost the same extent as S-1 alone (Fig. 3c). These results suggest that in the acto-S-1 rigor complex, F-actin binds with the S-site and, therefore, that the SH1 blocked sterically with F-actin cannot react with IAEDANS. In the presence of ATP, the S-site is probably dissociated from F-actin and the exposed SH1 may react freely with the reagent.

All these results are consistent with the hypothesis that during the steady state of ATPase, the S-site is dissociated

0 2 4 6 8 min

a) S-1

b) Acto-S-1

c) Acto-S-1
   +ATP

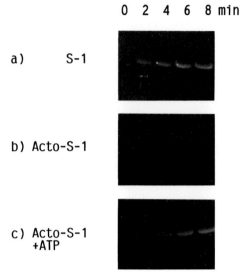

Fig. 3   Time course of the reaction between IAEDANS and S-1. Conditions were 10 mM KCl, 2 mM MgCl$_2$, 20 mM Tris-HCl (pH 8.0), 4.9 μM S-1 and 0.2 mM IAEDANS at 0 °C. Concentrations of ATP and F-actin were 2 mM and 20 μM, respectively. At the time indicated, the reaction was stopped by adding 2-mercaptoethanol to 10%, and the samples were applied to SDS-PAGE. The fluorescent band corresponding to the S-1 heavy chain was detected under a UV-light.

from F-actin, and the J-site maintains the weak acto-S-1 link. The question then arises as to how the dissociated S-site activates the ATPase activity.   To answer this question, it may be important to determine the reason why the Mg-ATPase activity of S-1 alone is so low.   It may be due to structural changes in the myosin head induced by ATP.   The structural change around the active site of ATPase has been shown by perturbations of Trp and Tyr chromophores [Morita, 1967, Werber et al., 1972].   Coupled structural changes also occur around the SH1 and SH2 groups.   The crosslinking reaction of the two SH groups by pPDM is enhanced in the presence of Mg-ATP [Reisler et al., 1974, 1977, Burke & Reisler, 1977, Wells & Yount, 1979].   The SH1 and the SH2 groups probably move into proximity during the ATPase reaction.   Furthermore, it has

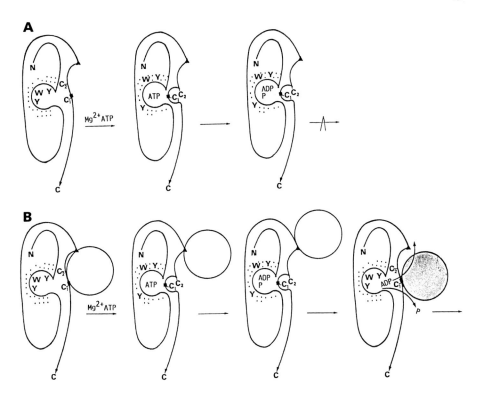

Fig 4. A proposed model of S-1 ATPase assuming that the S-site is an inhibitory site. The N- and C-terminals of the S-1 heavy chain are indicated by N and C, respectively; the actin binding S- and J-sites are indicated by ■ and ▲, respectively. (A) In the absence of F-actin. By the binding of Mg-ATP, the side chains of Trp (W) and Tyr (Y) around the active site of S-1 ATPase are buried inside, and the Cys(SH1) (C_1) and Cys(SH2) (C_2) move into mutual proximity. The trapped ATP at the active site is cleaved rapidly, and the products stay at the active site, suppressing the turnover rate of ATPase to a low level. (B) In the presence of F-actin. Both S- and J-sites on S-1 bind with an actin monomer (white circle) in F-actin in the absence of ATP. By the binding of Mg-ATP, the S-site is dissociated from the monomer and the affinity with F-actin is maintained at the weakly binding J-site. The dissociated S-site behaves similarly to that in (A) to trap Mg-ATP and to cleave ATP rapidly. When products are formed, the S-site re-binds transiently with F-actin, which rapidly releases the products, resulting in an increase in the turnover rate. If the freedom of the bound actin monomer increases after the dissociation of the S-site, and the re-bound actin monomer (shaded) is different from the one dissociated first, the sliding movement is possible. After releasing all products, the J-site also re-binds with the new actin monomer, and the ATPase cycle is repeated.

ATP-Binding Proteins	Residue Number from		
Myosin	698-	V L E G I R I C R K G F P S R I L Y A D F	1
Adenylate kinase	102-	G E E F E R K I - - G Q P T L L L Y V D A	2
Rho	246-	V I E K A K R L V E H K K D V I I L L D S	3
β-subunit of Mitochondrial ATPase	239-	V A E Y F R D Q - E G Q - D V L L F I D N	4

Table II   Homology of the peptide segment of the myosin heavy chain around the critical residues of the actin-binding S-site to segment B of the ATP-binding site. The critical residues for the actin binding are indicated by asterisks and Cys(SH1) is underlined. The conserved amino acid residues are boxed. The amino acid sequences are from 1, [Elzinga & Collins, 1977]; 2, [Hail et al., 1974]; 3, [Pinkham & Platt, 1983]; and 4, [Garboczi et al., 1988].

been observed that the nucleotides are trapped at the active site by the crosslinking reaction [Wells & Yount, 1979, Wells et al., 1979].   It seems highly probable that these structural changes of S-1 induced by the binding of Mg-ATP at the active site cause the high rate of ATP cleavage as well as the extremely low turnover rate of S-1 ATPase by trapping the ATPase products at the active site in the absence of F-actin. If this is the case, the actin-binding S-site including the SH1 may be considered to be an inhibitory site of S-1 ATPase. Kinetic studies show that the trapped products are acceleratedly released in the presence of F-actin [Lymn & Taylor, 1971, Tonomura, 1986].   F-actin, therefore, is assumed to deinhibit the inhibition by binding transiently with the inhibitory S-site.   The so called "actin-activation" of S-1 ATPase may be a kind of deinhibition.   These speculations are schematically shown in Fig. 4.

Finally, it should be pointed out that the actin-binding S-site may share a part of the ATP-binding site.   Many ATP-binding proteins contain two kinds of peptide segments, segment

A and segment B for the ATP-binding fold [Walker et al., 1982]. Segment A of myosin has been found in the N-terminal 27-kDa domain of S-1 heavy chain [Walker et al., 1982]. As shown in Table II, the amino acid sequence around the actin-binding S-site on the S-1 heavy chain was analogous to segment B of other ATP-binding proteins [Eto et al., 1990, Burke et al., 1990]. It is highly probable, therefore, that the dissociation of the S-site from F-actin in the presence of ATP is due to the competitive binding between ATP and F-actin at the S-site.

REFERENCES

Burke M, Reisler E (1977) Effect of nucleotide binding on the proximity of the essential sulfhydryl groups of myosin. Chemical probing of movement of residues during conformational transitions. Biochemistry 16: 5559-5563

Burke M, Rajasekharan N, Maruta S, Ikebe M (1990) A second consensus sequence of ATP-requiring proteins resides in the 21-kDa C-terminal segment of myosin subfragment-1. FEBS Lett. 262:185-188

Chaussepied P, Morales MF (1988) Modifying preselected sites on proteins: The stretch of residues 633-642 of the myosin heavy chain is part of the actin-binding site. Proc. Natl. Acad. Sci. USA 85:7471-7475

Elzinga M, Collins JH (1977) Amino acid sequence of a myosin fragment that contains SH-1, SH-2, and N-methylhistidine. Proc. Natl. Acad. Sci. USA 74: 4281-4284

Eto M, Suzuki R, Morita F, Kuwayama H, Nishi N, Tokura S (1990) Roles of the amino acid side chains in the actin-binding S-site of myosin heavy chain. J. Biochem. 108: 499-504

Garboczi DN, Fox AH, Gerring SL, Pedersen PL (1988) β-subunit of rat liver mitochondrial ATP synthase: cDNA cloning, amino acid sequence, expression in Escherichia coli, and structural relationship to adenylate kinase. Biochemistry 27: 553-560

Heil A, Muller G, Noda L, Pinder T, Schirmer H, Schirmer I, Von Zabern I (1974) The amino-acid sequence of porcine adenylate kinase from skeletal muscle. Eur. J. Biochem. 43: 131-144

Katoh T, Imae S, Morita F (1984) Binding of F-actin to a region between SH1 and SH2 groups of myosin subfragment-1 which may determine the high affinity of acto-subfragment-1 complex at rigor. J. Biochem. 95: 447-454

Katoh T, Morita F (1984) Interaction between myosin and F-actin. Correlation with actin-binding sites on subfragment-1. J. Biochem. 96:1223-1230

Katoh T, Katoh H, Morita F (1985) Actin-binding peptide obtained by the cyanogen bromide cleavage of the 20-kDa fragment of myosin subfragment-1. J. Biol. Chem. 260:6723-6727

Keane AM, Trayer IP, Levine BA, Zeugner C, Ruegg JC (1990) Peptide mimetics of an actin-binding site on myosin span two functional domains on actin. Nature 344:265-268

Lymn RW, Taylor EW (1971) Mechanism of adenosine triphosphate hydrolysis by actomyosin. Biochemistry 10: 4617-4624

Morita F (1967) Interaction of heavy meromyosin with substrate I. Difference in ultraviolet absorption spectrum between heavy meromyosin and its Michaelis-Menten complex. J. Biol. Chem. 242:4501-4506

Morita F, Suzuki R, Nishi N, Tokura S (1988) Synthetic actin-binding peptides having the amino acid sequence around the reactive cysteinyl residues (SH1) of myosin heavy chain. Pept. Chem. 1987: 743-746

Mornet D, Pantel P, Audemard E, Kassab R (1979) The limited tryptic cleavage of chymotryptic S-1: An approach to the characterization of the actin site in myosin heads. Biochem. Biophys. Res. Commun. 89: 925-932

Pinkham JL, Platt T (1983) The nucleotide sequence of the rho gene of E. coli K-12. Nucleic Acids Research 11: 3531-3545

Reisler E, Burke M, Himmelfarb S, Harrington WF (1974) Spatial proximity of the two essential sulfhydryl groups of myosin. Biochemistry 13: 3837-3840

Reisler E, Burke M, Harrington WF (1977) Reactivity of essential thiols of myosin. Chemical probes of the activated state. Biochemistry 16: 5187-5191

Sutoh K (1983) Mapping of actin-binding sites on the heavy chain of myosin subfragment 1. Biochemistry 22: 1579-1585

Suzuki R, Nishi N, Tokura S, Morita F (1987) F-actin-binding synthetic heptapeptide having the amino acid sequence around the SH1 cysteinyl residue of myosin. J. Biol. Chem. 262: 11410-11412

Suzuki R, Morita F, Nishi N, Tokura S (1990) Inhibition of actomyosin subfragment 1 ATPase activity by analog peptides of the actin-binding site around the Cys(SH1) of myosin heavy chain. J. Biol. Chem. 265: 4939-4943

Tonomura Y (1986) Energy-transducing ATPases--structure and kinetics. Cambridge University Press, Cambridge London New York New Rochelle Melbourne Sydney

Walker JE, Saraste M, Runswick MJ, Gay NJ (1982) Distantly related sequences in the a- and b-subunits of ATP synthase, myosin, kinases and other ATP-requiring enzymes and a common nucleotide binding fold. EMBO J. 1: 945-951

Wells JA, Yount RG (1979) Active site trapping of nucleotides by crosslinking two sulfhydryls in myosin subfragment 1. Proc. Natl. Acad. Sci. USA 76: 4966-4970

Wells JA, Knoeber C, Sheldon MC, Werber MM, Yount RG (1980) Crosslinking of myosin subfragment-1. Nucleotide-enhanced modification by a variety of bifunctional reagents. J. Biol. Chem. 255: 11135-11140

Werber MM, Szent-Gyorgyi AG, Fasman GD (1972) Fluorescence studies on heavy meromyosin-substrate interaction. Biochemistry 11: 2872-2883

Yamamoto K, Sekine T (1979) Interaction of myosin subfragment-1 with actin II. Location of the actin binding site in a fragment of subfragment-1 heavy chain. J. Biochem. 86:1863-1868

Competitive Inhibition of Maximum Ca-Activated Force in Skinned Muscle Fibers by Cationic Peptides from the SH-1 Region of Myosin Heavy Chain

P. Bryant Chase, Thomas W. Beck, John Bursell, and Martin J. Kushmerick*
Departments of Radiology and *Physiology and Biophysics
University of Washington, Seattle, Washington 98195, USA

Competition assays using peptides derived from the primary sequences of the contractile proteins provide a useful approach to identification of sites of protein-protein interaction and structure-function relationships of muscle proteins (Rüegg, 1990). There has been considerable interest in this method of studying the possible actin-binding role of the SH_1 (Cys 707) region of myosin heavy chain (Keane et al., 1990; Chase et al., 1990; Suzuki et al., 1987). The cationic nature of SH_1 peptides is implicated as the primary factor governing inhibition of force production by skinned fibers, and of MgATPase activity of acto-S1 (myosin subfragment 1) and myofibrils in solution; both assays involve cycling cross-bridges, but with different mechanical loading.

USE OF PEPTIDES IN SKINNED FIBER ASSAYS

Single muscle fibers with permeabilized membranes are a good test system for studying the force-generating interaction between actin and myosin. The experimenter is able to directly manipulate the environment of the contractile filaments and to regulate force development by altering the composition of the bathing solution under carefully controlled conditions. We chose to use peptides to probe the actomyosin interface because of their potential for specificity in binding to a complementary site as well as their potential for providing information about the equivalent sequence in the native molecule from which they were derived. As an alternative, antibodies have been used by others to block putative functional sites on contractile proteins (Harrington et al., 1990; Labbé et al., 1990; DasGupta and Reisler, 1989; Lovell et al., 1988; Miller et al., 1987; Méjean et al., 1986). However, antibodies only provide information about the site to which it is bound, i.e., the complementary binding site. More importantly, the relatively large size of an antibody (compare peptides on the order of 1 kD MW with Fab fragments of ~ 50 kD) results in an effectively lower spatial resolution view of the structure being examined and, in skinned fiber and myofibril assays, also results in a slower rate of diffusional equilibration throughout the myofilament lattice (Harrington et al., 1990; Lovell et al., 1988). Given the 2- to 10-fold lower apparent diffusion coefficient (D) of most molecules in muscle fibers compared with that in free solution (Maughan and Wegner, 1988; Kushmerick and Podolsky, 1969), we calculated that molecules on the order of 100 kD MW should equilibrate throughout a cylindrical muscle fiber of 100 μm diameter in a time period on the order of 10 min. However peptides of 1 kD MW should equilibrate on the time scale of 10 s to 1 min (Crank, 1975), which is consistent with the observed rate of change of maximum Ca-activated force following addition or removal of an inhibitory peptide.

EFFECTS OF NATIVE AND CONTROL SEQUENCE PEPTIDES

Our choice of peptides in the search for actin-binding sites on myosin was directed by the report that the SH_1 heptapeptide IRICRKG bound to F-actin and competitively inhibited rigor complex formation between S1 and F-actin in solution (Suzuki et al., 1987). This sequence is part of a highly conserved region in myosin heavy chain (Warrick and Spudich, 1987) which extends from Cys 697 to Pro 712 of rabbit skeletal muscle myosin heavy chain:

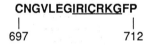

If such peptides block a part of actin that is necessary for force development, independent of whether the site is equivalent to a "ratchet" in a rocking cross-bridge model as envisioned by Huxley and Simmons (1971) or one of several portions of the primary sequence which act in concert, then force development should be blocked at least partly if not completely. Both the 7- (underlined) and 16-residue peptides inhibited maximum Ca-activated force in skinned muscle fibers with apparent $K_{0.5}$'s of 3.8 - 6 mM (Table 1; Chase et al., 1990). Blocking the end charges or adding additional conserved residues (i.e., enhancing the peptide's correspondence to the native molecule) did not increase its efficacy as an inhibitor.

During subsequent investigation of control sequences, we found that a scrambled sequence peptide (RICIRGK) was equally inhibitory (Table 1; Chase et al., 1990). Also, substitution of Pro for Cys in the native sequence heptapeptide did not alter its inhibitory properties (Chase et al., 1990). Additional control sequences demonstrated that the ability to inhibit maximum Ca-activated force was related to the net charge of the peptide (Table 1). A peptide with greater net positive charge was a more effective inhibitor while another peptide with lower net positive charge was a less effective inhibitor (Table 1). We concluded from these experiments that the mechanism by which SH_1 peptides inhibit maximum Ca-activated force of skinned fibers is related to their cationic nature. Further support for this conclusion was derived from data on the inhibition of acto-S1 ATPase in solution by SH_1 peptides (Keane et al., 1990; Suzuki et al., 1990). In these solution studies, variations in sequence were made by selecting different portions and lengths of the native myosin heavy chain sequence around SH_1; the apparent K_i was also correlated with net peptide charge, with the most positively charged peptides being the most inhibitory.

peptide	net charge	$K_{0.5}$ (mM)	s.e.
IRICRKG-NH$_2$	4+	2.5	0.1
IRICRKG	3+	3.8	0.2
Ac-IRICRKG-NH$_2$	3+	6.0	0.2
Ac-CNGVLEGIRICRKGFP-NH$_2$	3+	4.9	0.2
RICIRGK	3+	4.1	0.1
IEICRKG	1+	>> 7 (27)[a]	(7)

[a] Numbers in parenthesis () are estimates based on extrapolation from the concentrations tested.

Table 1. Apparent inhibitory constants ($K_{0.5}$'s) for SH_1 peptides derived from inhibition of maximum Ca-activated force (pCa 4.5) of single skinned fibers from rabbit psoas muscle. Experiments were performed as described elsewhere (Chase and Kushmerick, 1988; Chase PB, Beck TW, Bursell J, Kushmerick MJ, manuscript submitted).

DISCUSSION

Our data do not support an actin-binding role for the SH_1 portion of myosin heavy chain during force development by cross-bridges. The proposed role for the SH_1 region in modulation of the thin filament's response to Ca^{2+} (Keane et al., 1990) should also be examined in more detail for evidence of sequence specificity. In contrast, inhibition of rigor complex formation was found to be more sensitive to the sequence of SH_1 peptides than to their net charge (Eto et al., 1990; Suzuki et al., 1987). This difference between cycling cross-bridges and those in rigor appears to be in accord with previous conclusions that ionic interactions are less important for rigor binding than for actomyosin MgATPase activity (Katoh and Morita, 1984; Sutoh, 1983; Gordon et al., 1973).

Two other classes of experiments have been cited as support for an actin-binding role of the SH_1 region of myosin. First, the rate of SH_1 - SH_2 (Cys 697) crosslinking by N,N'-p-phenylenedimaleimide (pPDM) was decreased by actin added in rigor conditions (Katoh et al., 1984). However, in the absence of actin, the crosslinking rate was increased by ATP, which binds to S1 at a site distant from SH_1 (Tokunaga et al., 1987; Sutoh et al., 1984). These results suggest that, rather than a steric blocking of SH_1 and SH_2 by actin, ATP and actin have opposite, allosteric effects on the global structure of S1. The second class of experiments utilized NMR to detect the effects of actin-binding by S1 which had been covalently modified at SH_1. Decreased ^1H spectral intensity of some actin residues in the presence of a 4-maleimido-2,2,6,6-tetramethylpiperidinooxyl paramagnetic spin label (Keane et al., 1990) was illustrated as occuring in residues which are distributed throughout the primary structure of actin peptide 1 - 44. In combination with the inherent difficulties associated with obtaining narrow-line ^1H spectra from large proteins, this suggests that detection of specific, functionally significant interactions may be difficult with this approach. An actin induced immobilization of a fluorinated label ((trifluoromethyl)mercury; Barden et al., 1989) could be explained by any of three classes of mechanisms: direct interaction of the label with actin, as suggested by the authors; an allosteric immobilization of the SH_1 region of S1 upon actin binding; or interactions between neighboring S1's under conditions where most of the sites on actin have S1 bound.

Thus, we believe that the evidence for an actin-binding role of the SH_1 region of myosin during active cross-bridge cycling is not compelling, although this highly conserved portion of the primary sequence is likely to have an important role in energy transduction by actomyosin. A concensus nucleotide-binding site has been noted near SH_1 (Eto et al., 1990; Burke et al., 1990), but the functional significance of a second ATP-binding site in the myosin head is not yet obvious. The SH_1 region has recently been proposed to take part in linking the ATP site and the actin site in S1 (Eto et al., 1990; Botts et al., 1989), although considerable work will be required to verify this hypothesis. Thus the function of the SH_1 region remains to be elucidated.

FUTURE DIRECTIONS

Other putative actin-binding regions of myosin also have high densities of basic residues (Eldin et al., 1990; Chaussepied and Morales, 1988; Sutoh, 1983; Mornet et al., 1981a,b). Therefore further studies involving competition by peptides from these regions should be designed to distinguish between effects due to the charge of cationic peptides and sequence-specific effects. The actin-binding capacity of cationic peptides (Suzuki et al., 1987) may make them useful probes of the myosin-binding site on actin, providing complementary information to an anti-actin antibody which, like the heptapeptide, inhibits actin-activation of S1 MgATPase activity but not rigor binding (DasGupta and Reisler, 1989; Miller et al., 1987).

ACKNOWLEDGEMENTS

Supported by the US National Institutes of Health (Grants HL 31962 and AM 36281).

REFERENCES

Barden JA, Phillips L, Cornell BA, dos Remedios CG (1989) ^{19}F NMR studies of the interaction of selectively labeled actin and myosin. Biochemistry. 28:5895-5901

Botts J, Thomason JF, Morales MF (1989) On the origin and transmission of force in actomyosin subfragment 1. Proc. Natl. Acad. Sci. USA 86:2204-2208

Burke M, Rajasekharan KN, Maruta S, Ikebe M (1990) A second consensus sequence of ATP-requiring proteins resides in the 21-kDa C-terminal segment of myosin subfragment 1. FEBS Lett. 262:185-188

Chase PB, Kushmerick MJ (1988) Effects of pH on contraction of rabbit fast and slow skeletal muscle fibers. Biophys. J. 53:935-946

Chase PB, Bursell J, Kushmerick MJ (1990) Peptide inhibition of rabbit psoas muscle skinned fiber mechanics & myofibrillar ATPase activity. Biophys. J. 57:538a

Chaussepied P, Morales MF (1988) Modifying preselected sites on proteins: the stretch of residues 633-642 of the myosin heavy chain is part of the actin-binding site. Proc. Natl. Acad. Sci. USA 85:7471-7475

Crank J (1975) The Mathematics of Diffusion, 2nd ed. Oxford Univ. Press, Oxford

DasGupta G, Reisler E (1989) Antibody against the amino terminus of α-actin inhibits actomyosin interactions in the presence of ATP. J. Mol. Biol. 207:833-836

Eldin P, Le Cunff M, Diederich KW, Jaenicke T, Cornillon B, Mornet D, Vosberg H-P, Leger JJ (1990) Expression of human β-myosin heavy chain fragments in *Escherichia coli*; Localization of actin interfaces on cardiac myosin. J. Muscle Res. Cell Motil. 11:378-391

Eto M, Suzuki R, Morita F, Kuwayama H, Nishi N, Tokura S (1990) Roles of the amino acid side chains in the actin-binding S-site of myosin heavy chain. J. Biochem. 108:499-504

Gordon AM, Godt RE, Donaldson SK, Harris CE (1973) Tension in skinned frog muscle fibers in solutions of varying ionic strength and neutral salt composition. J. Gen. Physiol. 62:550-574

Harrington WF, Karr T, Busa WB, Lovell SJ (1990) Contraction of myofibrils in the presence of antibodies to myosin subfragment 2. Proc. Natl. Acad. Sci. USA 87:7453-7456

Huxley AF, Simmons RM (1971) Proposed mechanism of force generation in striated muscle. Nature 233:533-538

Katoh T, Morita F (1984) Interaction between myosin and F-actin. Correlation with actin-binding sites on subfragment-1. J. Biochem. 96:1223-1230

Katoh T, Imae S, Morita F (1984) Binding of F-actin to a region between SH_1 and SH_2 groups of myosin subfragment-1 which may determine the high affinity of acto-subfragment-1 complex at rigor. J. Biochem. 95:447-454

Keane AM, Trayer IP, Levine BA, Zeugner C, Rüegg JC (1990) Peptide mimetics of an actin-binding site on myosin span two functional domains on actin. Nature 344:265-268

Kushmerick MJ, Podolsky RJ (1969) Ionic mobility in muscle cells. Science 166:1297-1298

Labbé J-P, Méjean C, Benyamin Y, Roustan C (1990) Characterization of an actin-myosin head interface in the 40-113 region of actin using specific antibodies as probes. Biochem. J. 271:407-413

Lovell S, Karr T, Harrington WF (1988) Suppression of contractile force in muscle fibers by antibody to myosin subfragment 2. Proc. Natl. Acad. Sci. USA 85:1849-1853

Maughan D, Wegner E (1988) Diffusion coefficients of cytosolic proteins in rabbit muscle. Biophys. J. 53:61a

Méjean C, Boyer M, Labbé JP, Derancourt J, Benyamin Y, Roustan C (1986) Antigenic probes locate the myosin subfragment 1 interaction site on the N-terminal part of actin. Biosci. Rep. 6:493-499

Miller L, Kalnoski M, Yunossi Z, Bulinski JC, Reisler E (1987) Antibodies directed against N-terminal residues on actin do not block acto-myosin binding. Biochemistry 26:6064-6070

Mornet D, Bertrand R, Pantel P, Audemard E, Kassab R (1981a) Structure of the actin-myosin interface. Nature 292:301-306

Mornet D, Bertrand R, Pantel P, Audemard E, Kassab R (1981b) Proteolytic approach to structure and function of actin recognition site in myosin heads. Biochemistry 20:2110-2120

Rüegg JC (1990) Muscle protein interaction; Competition by peptide mimetics. J. Muscle Res. Cell Motil. 11:189-190

Sutoh K (1983) Mapping of actin-binding sites on the heavy chain of myosin subfragment 1. Biochemistry 22:1579-1585

Sutoh K, Yamamoto K, Wakabayashi T (1984) Electron microscopic visualization of the SH_1 thiol of myosin by the use of an avidin-biotin system. J. Mol. Biol. 178:323-339

Suzuki R, Morita F, Nishi N, Tokura S (1990) Inhibition of actomyosin subfragment 1 ATPase activity by analog peptides of the actin-binding site around the Cys(SH1) of myosin heavy chain. J. Biol. Chem. 265:4939-4943

Suzuki R, Nishi N, Tokura S, Morita F (1987) F-actin-binding synthetic heptapeptide having the amino acid sequence around the SH1 cysteinyl residue of myosin. J. Biol. Chem. 262:11410-11412

Tokunaga M, Sutoh K, Toyoshima C, Wakabayashi T (1987) Location of the ATPase site of myosin determined by three-dimensional electron microscopy. Nature 329:635-638.

Warrick HM, Spudich JA (1987) Myosin structure and function in cell motility. Ann. Rev. Cell Biol. 3:379-421

The Use of Peptide Mimetics to Define the Actin-Binding Sites on the Head of the Myosin Molecule

Ian P Trayer, Anita M Keane, Zeki Murad, J Caspar Rüegg* and K John Smith
School of Biochemistry, The University of Birmingham, Edgbaston, Birmingham B15 2TT, UK and *II Physiologisches Institut, Universitat Heidelberg, Im Neuerheimer Feld 326, 6900 Heidelberg, Germany

Definition of the molecular mechanisms in muscle contraction and its regulation entails a description of how the components of the organised assembly of proteins first dock with their substrates/partners, then interact and transmit information through the molecular array. Crosslinking studies and experiments with proteolysed fragments of the myosin head (subfragment 1, S1) from a variety of laboratories [e.g. Chaussepied et al., 1986a; Sutoh, 1983] have indicated the approximate regions of the molecule involved in complex formation. In order to define precisely the exact locations of these interfaces we have synthesized peptides based on the S1 sequence and tested these for their ability to bind to actin and influence its biological properties. Such chemical synthesis allows small regions of the parent protein, usually not obtainable by proteolytic or chemical digestion, to be examined in isolation. It is not expected *a priori* that these fragments will adopt the same structure in solution as the parent protein, although many NMR studies of late are reporting that small peptides in solution are adopting distinct conformational preferences, but rather that they contain sufficient 'information' to recognise specifically their binding site when presented with a suitable molecular template, *viz* actin. Such a peptide mimetic approach can be used to delineate the extent of the molecular interfaces and determine their structure by NMR methods and to probe the mechanism of action of the host system in solution enzymic assays and in skinned muscle fibres. Using this approach we have so far identified three sites on the myosin head that interact with actin and these are described below.

(i) The Site on the 20kDa Domain of S1 around Cysteine 707

The region around the fast-reacting SH$_1$ thiol (C$_{707}$) is heavily conserved between myosins from various species. Our attention focused on this segment when it was observed that a spin-label attached to C$_{707}$ specifically broadened resonances from the N-terminal region of actin in NMR studies [Keane et al., 1990], indicating the proximity of the regions from the two molecules in the acto-S1 complex. A variety of peptides from this region were synthesized (Fig.1) and those containing the region 695-727 were found to bind to actin both in NMR experiments and by direct binding studies.

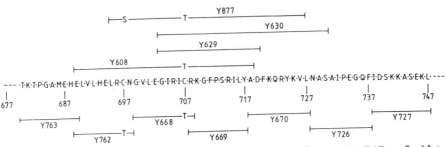

Fig.1 Peptides synthesized from the region 677-747 of the 20kDa domain of S1 are of lengths indicated by the bars. Where indicated, the SH$_1$ thiol (C$_{707}$) and the SH$_2$ thiol (C$_{697}$) have been replaced by S or T. All peptides were purified by HPLC.

The NMR data emphasized that only selected amino acids and groups of amino acids along this sequence were involved in direct contact with actin. For example, in peptide Y670, binding was confined predominantly to Y$_{724}$, V$_{726}$ and L$_{727}$. Peptides incorporating this region also acted as mixed inhibitors of the actin-stimulated MgATPase of S1 and influenced the contractile force developed in skinned fibres [Keane et al., 1990], whereas peptides flanking this region did not bind to actin in either these *in vitro* or *in vivo* experiments and so acted as controls. The K$_i$ values derived from the enzyme assays were in close agreement with the K$_d$ values found in direct binding experiments. Morita and her colleagues [Suzuki et al., 1987; 1990] have also reported that a small peptide from this region (704-710) interfers with S1 binding to actin.

Several interesting features emerged from the biological assays of these peptides: (a) All peptides from this region inhibited force development at maximal activating $[Ca^{2+}]$'s, i.e. the strong binding state [Keane et al., 1990]. (b) Peptides from the N-terminal end of this region acted as Ca-sensitizers by increasing force development at submaximal activating $[Ca^{2+}]$'s, whereas peptides from or containing the C-terminal 10 residues inhibited force development at all $[Ca^{2+}]$'s [Keane et al., 1990; Rüegg et al., 1991]. This is displayed graphically in the form of an activity profile in Fig.2.

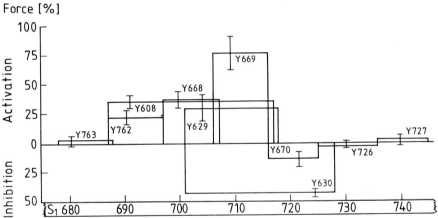

Fig.2 Activity profile of synthetic peptides. The peptide numbers correspond to those in Fig.1 and the residue numbers for the S1 sequence form the abscissa. Activation or inhibition of force development in rabbit psoas muscle skinned fibres (the ordinate) is indicated as per cent of control contraction before addition of 50μM peptide elicited at submaximal activating concentrations of Ca^{2+} (pCa = 5.6). (Average ± s.e. of 5-6 experiments). Details of fibre preparations are given in Keane et al. (1990).

These different responses from different segments of this region suggest that it is binding to two functionally distinct sites on actin that may well be spatially distinct as well. This opens up various possibilities for interaction that are depicted schematically in Fig.3.

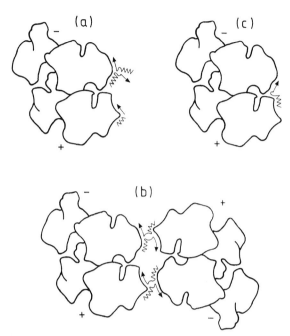

Fig.3 Schematic respresentation of peptides from the 20kDa domain binding at two distinct sites on the F-actin filament. (a) Shorter peptides (∿∿ and ⟶) bind 1 mol per mol actin monomer, whereas longer peptides (∿∿⟶) may bind 2 mol per mol. (b) Longer peptides can therefore bundle actin filaments. (c) Longer peptides may bind across two actin monomers.

In Fig.3a, it is shown that two mol peptide containing the whole region (e.g. Y630, Y877) may bind to one actin monomer in the filament, whereas the "half peptides", containing only one site, will only bind one mol per mol actin. This is indeed what is found in binding studies where, for example, in Scatchard plots n=2 for Y630 but n=1 for Y629. This is confirmed by noting that the longer peptides are capable of bundling actin filaments in electronmicroscope studies whereas the shorter ones are not (see Fig.3b). If this region of S1 is capable of spanning two actin monomers in the filament (Fig.3c) rather than one, then the above will also be true but in addition, crosslinking the peptides to actin will produce actin polymers on SDS-polyacrylamide gels as well as monomers and dimers. Crosslinking Y877 to actin by a sulpho-N-hydroxysuccinimide-enhanced carbodiimide reaction produced this result.   Thus, it is possible that this region in S1 is capable of spanning two actin monomers in the actin filament; an observation that finds some support in recent image reconstruction analyses of thin filaments obtained by cryo-electron microscopy [Milligan et al., 1990].

(ii) **The Site on the C-terminal Region of the   50kDa Domain of S1**

The crosslinking studies of Sutoh [Sutoh, 1983] suggested that a binding site for actin existed within the C-terminal 8kDa region of the 50kDa domain of S1.   This is supported by observations that renatured 30kDa [Chaussepied et al., 1986b] and 26kDa [Griffiths and Trayer, 1989] fragments of S1, containing the whole of the 20kDa domain and either some 10kDa or 6kDa C-terminal portions of the 50kDa domain, bound to actin in an ATP-sensitive manner whereas the acto-20kDa domain complex was not dissociated by ATP [Chaussepied et al., 1986b].   Several peptides were therefore synthesized spanning the region 560-616 of the C-terminal end of the 50kDa domain. When tested by NMR, only the C-terminal peptide from this region (592-616) bound to actin, and here binding was confined to the C-terminus (Fig.4).   Binding to actin appears to be confined to residues 610-616 and may well extend C-terminal to this.   This site, therefore, remains to be delineated fully, but it should be noted that it is very close to the 'loop' region connecting the 50kDa and 20kDa domains of S1.   This K,G-rich loop has been suggested as an actin-binding site [Chaussepied and Morales, 1988] based upon the properties of an S1 species, in which a charge-matching 'antipeptide' has been crosslinked to this loop and interfers with acto-S1 interaction.

(iii)   **The Site on the N-terminal Region of the Alkali 1 (LC1) Myosin Light Chain**

All vertebrate striated muscle myosins contain at least one alkali 1 (A1)-type light chain.   These A1-light chains are characterized by containing an N-terminal extension of 30-40

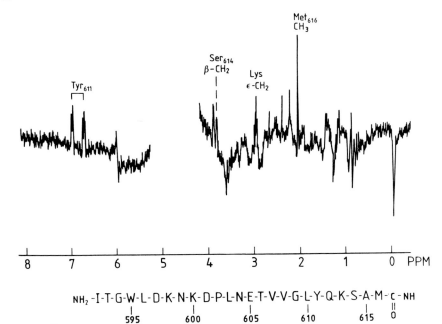

Fig.4 ^{1}H-NMR difference spectrum showing side chain residues of myosin 50kDa domain peptide 592-616 perturbed by actin. "Marker" resonances deriving from amino acids found only at the C-terminal end of the peptide are indicated on the spectrum (Y_{611}, S_{614} and M_{616}). Note that other amino acid occurring only once in the sequence, e.g. W_{595} and P_{602}, do not appear in the difference spectrum.

amino acids whose sequence is (a) dominated by extensive $(X-P)_n$ tandem repeat motifs [Grand, 1982] and (b) terminates in a fully methylated N-terminal amino acid, usually α–N-trimethyl-alanine (Me$_3$A) [Henry et al, 1985]. The N-terminal region of these light chains has been found to bind to the C-terminal region of actin in both NMR [Prince et al, 1981; Trayer et al., 1987] and crosslinking studies [Sutoh, 1982]. This region displays a high degree of segmental mobility relative to the rest of the S1 molecule and dominates the NMR spectrum of S1(A1). The $(X-P)_n$ repeat motif (in rabbit fast skeletal muscle this is $(A-P)_7$) confers a rigid, extended structure to the region [Bhandari et al., 1986] that acts as a 'spacer' segment linking the actin binding site to the body of the light chain associated with the myosin head. Such sequence-constrained $(X-P)_n$ repeats are found in all A1-

type light chains and in numerous other proteins where they may also serve as a rigid spacer connecting functional domains. For example, in the TonB protein of E.coli, two $(X-P)_6$ regions (where X largely equals K in one case and E in the other) appear to span the periplasmic space linking the inner and outer membranes [Brewer et al., 1990].

We have recently synthesized several peptides from this region, some terminating in $(Me)_3A$ and some with the α-amino group free (Fig.5a) and monitored their binding to actin by NMR. Both the whole region (1-32) and the N-terminal 10 residues bind to actin whereas peptides 1-3 and 8-31 do not. Comparison of the binding of 1-32 and 1-10 suggests that the whole of the actin-binding region is confined to the first 8 or 9 amino acids, but surprisingly that the N-terminal methylation is not essential for binding (Fig.5c,d) (note that the $\beta-CH_3$ of A_1 is broadened in each case).

It has been known for some time that S1(A1) binds to regulated actin in the presence of Ca^{2+} more tightly than S1(A2) [Trayer and Trayer, 1985] and so this region may play a role in modulating the regulatory response. This is especially so since it was noted that A1 light chain binding to actin weakens the actin-Tn-I interaction [Grand et al., 1983].

Concluding Remarks

This article reports on at least three sites on S1 that interact with actin. Other sites may exist, for example, Mulrad (1989) has suggested involvement of the 27kDa domain, or may be part of the sites described here, for example, the 50kDa-20kDa loop region [Chaussepied and Morales, 1988] which is only 15 amino acids from the 50kDa site described here. At this moment it is not possible to determine how far apart these regions are on the S1 surface, it may be that they are relatively closely clustered together in space. Nevertheless, it does suggest that actin and myosin dock at multiple sites, which offers some interesting possibilities for the definition of weak and strong binding states (i.e. pre- and post-power stroke) of the actomyosin complex. It can readily be

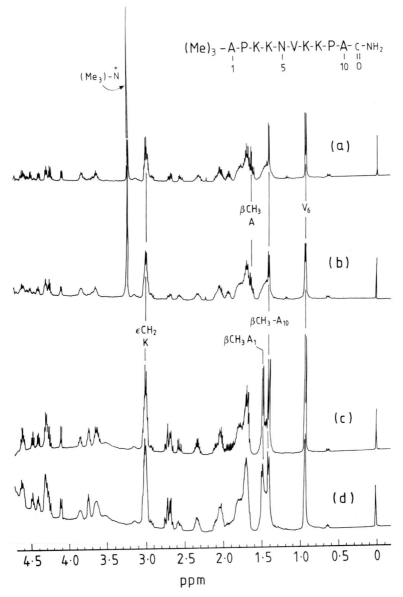

Fig.5 Binding of actin to the N-terminal 10 residues of the Al light chain. Two 10 residue peptides were synthesized, one terminating in α–amino-A ((c) and (d)) and one terminating in (Me)₃A ((a) and (b)). The ¹H-NMR spectra of the peptides alone are shown in (a) and (c) and after addition of actin (peptide:actin, 15:1) in (b) and (c).

envisaged that ATP binding can modulate the number of interprotein contact sites. The differential effects of the S1-derived peptides on muscle mechanics [Kraft et al. 1991, Rüegg et al. 1991] would suggest this to be so. Thus, all three sites would be involved in actin binding in the rigor-like complex, but in the weak binding state the presence of ATP could reduce the number of actin contact sites. This multidocking scenario appears to be widespread in the contractile machinery with troponin-I and troponin-C and troponin-I and actin each sharing two interaction sites [Grand et al, 1982; Levine et al, 1988] which could also be modulated as part of the contractile cycle.

There are several reports of differently-located myosin binding sites on actin which support this multidocking concept [for example, Trayer et al, 1987; Levine et al, 1990]. Many of these are located in subdomain 1 of the actin structure [Kabsch et al., 1990] and this appears to agree with recent image-reconstruction studies [Milligan et al., 1990]. This latter work also suggests that a single S1 molecule can make contact with two adjacent actin monomers in the longpitch helix, which is consistent with some of the data presented here and with the observation the S1 binds to actin peptides around residue 40 [Moir et al., 1987]. This region of actin is at the top of subdomain 2 and is proposed to be close to subdomain 1 of the next actin monomer along in the helix [Holmes et al., 1990]. The idea of the myosin head binding to two actin monomers in the rigor complex is appealing since it automatically angles the head relative to the actin helix. A weak binding state which may only bind to one actin monomer would then be free to adopt a different angle of attachment, much as most of the current models of contraction predict.

The idea of sharing/overlapping sites on actin for various actin-binding proteins is steadily emerging from the literature. Given the relatively restricted area on the outermost N-terminal domain of the actin monomer unit in the filament and the large number of proteins to which it binds this is hardly surprising. Subdomain 1, for example, is known to bind to S1, troponin-I, caldesmon, depactin and gelsolin.

This, in turn, implies that some rather interesting features of the molecular recognition processes must emerge in future whereby the same region on actin can invoke different molecular signal transduction pathways in different partners and possibly *vice versa*.

Finally, it only remains to emphasize the usefulness of the peptide approach in locating cognitive surfaces between interacting partner proteins and in probing the mechanism of their interaction. In future, NMR methods will be used to determine the absolute configuration of the peptides bound to their partner in order to fully understand the requirements of recognition process and perhaps suggest how selective intervention may be achieved.

References

Bhandari, D.G., Levine, B.A., Trayer, I.P. & Yeadon, M.E. (1986) ^{1}H-NMR study of mobility and confirmational constraints within the proline-rich N-terminal of the LC1 alkaline light chain of skeletal myosin. Correlation with similar segments in other protein systems. Eur. J. Biochem. 160: 349-356.

Brewer, S., Tolley, M., Trayer, I.P., Barr, G.C., Dorman, C.J., Higgins, C.F., Evans, J.S., Levine, B.A. & Wormald, M.R. (1990) Structure and function of X-Pro dipeptide repeats in the TonB protein from *S.typhimurium* and *E.coli*. J. Mol. Biol. 216: 881-893.

Chaussepied, P. & Morales, M.F. (1988) Modifying preselected sites on proteins: the stretch of residues 633-642 of the myosin heavy chain is part of the actin-binding site. Proc. Natl. Acad. Sci. USA 85: 7471-7475.

Chaussepied, P., Mornet, D., Audemard, E., Kassab, R., Goodearl, A.J., Levine, B.A. & Trayer, I.P. (1986a) Properties of the alkali light chain-20 kilodalton fragment complex from skeletal myosin heads. Biochemistry, 25: 4540-4547.

Chaussepied, P., Mornet, D. & Kassab, R. (1986b) Identification of polyphosphate recognition sites communicating with actin sites on skeletal myosin S1. Biochemistry, 25: 6426-6432.

Grand, R.J.A. (1982) The structure and function of myosin light chains. Life Sci. Reports, 1: 105-160.

Grand, R.J.A., Henry, G.D., Moir, A., Perry, S.V., Trayer, I.P., Dalgarno, D.C., Levine, B.A. & Parker, S.B. (1983) Modulation by troponin-C of the troponin-I inhibition of skeletal actomyosin interaction. A PMR spectral study. In de Bernard, B., Sottocasa, G.L., Sandri, G., Carafoli, E., Taylor, A.N., Vanaman, T.C. & Williams, R.J.P. (eds): Calcium-binding proteins, 1983. Elsevier Science Publishers, Amsterdam, pp 379-380.

Grand, R.J.A., Levine, B.A. & Perry, S.V. (1982) Proton-magnetic-resonance studies on the interaction of rabbit skeletal-muscle troponin I with troponin C and actin. Biochem. J. 203: 61-68.

Griffiths, A.J. & Trayer, I.P. (1989) Selective cleavage of skeletal myosin subfragment 1 to form a 26kDa peptide which shows ATP sensitive actin binding. FEBS Lett. 242: 275-278.

Henry, G.D., Trayer, I.P., Brewer, S. & Levine, B.A. (1985) The widespread distribution of α-N-trimethylalanine as the N-terminal amino acid of light chains from vertebrate striated muscle myosins. Eur. J. Biochem. 148: 75-82.

Holmes, K.C., Popp, D., Gebhard, W. & Kabsch, W. (1990) Atomic model of the actin filament. Nature, 347: 44-49.

Kabsch, W., Mannherz, H.G., Suck, D., Pai, E. & Holmes, K.C. (1990) Atomic structure of the actin: DNase I complex. Nature 347: 37-44.

Keane, A.M., Trayer, I.P., Levine, B.A., Zeugner, C. & Rüegg, J.C. (1990) Peptide mimetics of an actin-binding site on myosin span two functional domains on actin. Nature, 344: 265-268.

Kraft, Th., Trayer, I.P. & Brenner, B. (1991) Myosin peptides and cross-bridge kinetics. In Rüegg, J.C. (ed): Peptides as probes in muscle research. Springer-Verlag, Heidelberg, (this volume).

Levine, B.A., Moir, A.J.G. & Perry, S.V. (1988) The interaction of troponin-I with the N-terminal region of actin. Eur. J. Biochem. 172: 389-397.

Levine, B.A., Moir, A.J.G., Trayer, I.P. & Williams, R.J.P. (1990) Nuclear magnetic resonance studies of calcium modulated proteins and actin-myosin interaction. In Squire, J. (ed): Molecular mechanisms in muscle contraction. McMillan Press, London. pp 171-209.

Milligan, R.A., Whittaker, M. & Safer, D.(1990) Molecular structure of F-actin and location of surface binding sites. Nature, 348: 217-221.

Moir, A.J.G., Levine, B.A., Goodearl, A.J. & Trayer, I.P. (1987) The interaction of actin with myosin subfragment 1 and with pPDM-crosslinked S1: a ^{1}H-NMR investigation. J. Muscle Res. & Cell Motil. 8: 68-69.

Muhlrad, A. (1989) Isolation and characterization of the N-terminal 23-kilo Dalton fragment of myosin subfragment 1. Biochemistry, 28: 4002-4010.

Prince, H.P., Trayer, H.R., Henry, G.D., Trayer, I.P., Dalgarno, D.C., Levine, B.A., Cary, P.D. & Turner, C. (1981) Proton nmr spectroscopy of myosin subfragment-1 isoenzymes. Eur. J. Biochem. 121: 213-219.

Rüegg, J.C., Van Eyk, J., Hodges, R. & Trayer, I.P. (1991) Myosin and troponin peptides affecting Ca^{2+}-sensitivity of skinned fibres. In Rüegg, J.C. (ed): Peptides as probes in muscle research. Springer-Verlag, Heidelberg, (this volume).

Sutoh, K. (1982) Identification of myosin-binding sites on the actin sequence. Biochemistry, 21: 3654-3661.

Sutoh, K. (1983) Mapping of actin-binding sites on the heavy chain of myosin subfragment 1. Biochemistry, 22: 1579-1585.

68

Suzuki, R., Nishi, N., Tokura, S. & Morita, F. (1987) F-actin binding synthetic heptapeptide having the amino acid sequence around the SH1 cysteinyl residue of myosin. J. Biol. Chem. 262: 11410-11412.

Suzuki, R., Morita, F., Nishi, N. & Tokura, S. (1990) Inhibition of actomyosin subfragment 1 ATPase activity by analog peptides of the actin-binding site around the Cys(SH1) of myosin heavy chain. J. Biol. Chem. 265: 4939-4943.

Trayer, H.R. & Trayer, I.P. (1985) Differential binding of rabbit fast muscle myosin light chain isoenzymes to regulated actin. FEBS Lett. 180: 170-174.

Trayer, I.P., Trayer, H.R. & Levine, B.A. (1987) Evidence that the N-terminal region of the Al-light chain of myosin interacts directly with the C-terminal region of actin; a proton magnetic resonance study. Eur. J. Biochem. 164: 259-266.

Interference of Myosin Peptides with Weak and Strong Actin Interaction of Cross-Bridges in Skeletal Muscle Fibres

T. Kraft[*], U. Rommel[*], I.P. Trayer[#] and B. Brenner[*]
[*]Department of General Physiology, University of Ulm, Albert-Einstein-Allee 11, D7900 Ulm, FRG
and [#]School of Biochemistry, University of Birmingham, Birmingham B15 2TT, UK.

It is generally accepted that muscle contraction is the result of cyclic interaction of parts of the myosin molecules (the myosin heads, or the "cross-bridges") with the actin filaments (A.F. Huxley, 1957, 1974; Pringle, 1967; H.E.Huxley, 1969). According to recent biochemical studies (Stein et al., 1979; Rosenfeld and Taylor, 1984; Hibberd and Trentham, 1986) and mechanical and X-ray diffraction experiments on skinned fibers (Brenner et al., 1982, 1984, 1986; Yu and Brenner, 1989), cross-bridges exist in two main configurations, a configuration with low actin affinity (weak cross-bridge binding) and a configuration with high actin affinity (strong cross-bridge binding). During ATP-hydrolysis, cross-bridges are thought to cycle between these two main configurations (Eisenberg and Greene, 1980; Eisenberg and Hill, 1985). Furthermore, it was proposed that force is generated when the attached cross-bridge changes its configuration from the weak- to the strong-binding form (Eisenberg and Hill, 1985).

Recent equatorial X-ray diffraction studies revealed a difference in the mode of weak and strong cross-bridge interaction with actin (Yu and Brenner, 1989; Yu et al., 1990).

Similarly, from electron microscopy it was suggested that the mode of interaction between actin and S-1 is different for rigor and relaxing conditions (Craig et al., 1985). Based on these findings it was proposed that the sites of interaction between actin and myosin may be different for the weak and strong cross-bridge attachment to actin.

We tested this hypothesis by incubating membrane free (skinned) fibers of the rabbit psoas muscle with synthetic oligopeptides (8-10 amino acid residues) to see whether some of these peptides can specifically interfere with either the weak or the strong cross-bridge interaction. Here we describe experiments with peptides which represent parts of the primary sequence of the S1 molecule ("myosin" peptides). One of the tested peptides appears to interfere specifically with the weak cross-bridge interaction while the strong interaction is unaffected.

All experiments were performed on single fibers from the rabbit psoas muscle which were prepared and chemically skinned with Triton X-100 according to Brenner (1983) and Yu and Brenner (1989). Experimental solutions: Relaxing solution contained 1mM EGTA, 2mM $MgCl_2$, 1mM MgATP, 10mM imidazole, 2mM dithiothreitol. Preactivating and activating solution contained 2mM $MgCl_2$, 1mM MgATP, 10mM creatine phosphate, 200-250U/ml creatine kinase, 10mM caffeine, 10mM imidazole, 2mM dithiothreitol, and 1mM EGTA or 1mM CaEGTA respectively. The pH of all solutions was adjusted to 7.0 at the experimental temperature (5°C). Ionic strength was set to 170mM by adding the appropriate amount of K-propionate, or creatine phosphate.

The peptides were synthesized and purified by J.P. Trayer as described earlier (Keane et al., 1990). The purity of all peptides was checked by analytical HPLC and by amino-acid analysis. The binding of the peptides to actin was investigated in direct binding assays (Keane et al., 1990). The peptides were further characterized with respect to their inhibition constant of the actin-S1-Mg^{++}-ATPase activity (Keane et al., 1990).

The mechanical apparatus, including laser light diffraction was previously described (Brenner, 1980; Brenner and Eisenberg, 1986). The equipment for equatorial X-ray diffraction was described by Yu and Brenner (1989).

Fiber stiffness was measured by imposing ramp-shaped stretches to one end of the fibers. Stretch velocity was varied between 10^{-1} and 5×10^5 (nm/half-sarcomere)s^{-1}, and force was displayed vs. change in sarcomere length on a digital storage oscilloscope. Fiber stiffness was defined as the ratio of force change (ΔF) over the imposed filament sliding ($\Delta SL/2$) for the first 1nm/half-sarcomere of the applied stretch (chord stiffness; Simmons and Jewell, 1974).

Isometric force together with the time course of force redevelopment were recorded according to the method described previously (Brenner, 1985).

EFFECTS OF MYOSIN PEPTIDES UPON WEAK CROSS-BRIDGE BINDING TO ACTIN

Interference of myosin peptides with weak cross-bridge attachment to actin was probed by (i) measurements of fiber stiffness under relaxing conditions at low ionic strength (Brenner et al., 1982, 1986), and (ii) by recording equatorial X-ray diffraction patterns under the same conditions (Brenner et al., 1984). Using caldesmon and its actin binding fragments, we previously demonstrated that inhibition of weak cross-bridge attachment to actin results in reduction of stiffness and of the intensity ratio I_{11}/I_{10} of the two innermost equatorial reflections in relaxed fibers (Brenner et al., 1990; Kraft et al., 1991).

As shown in Fig. 1, addition of peptide Y668 (residues 697-706 of the 20kDa domain; Keane et al., 1990) results in a significant decrease in relaxed fiber stiffness. However, no significant effect was found for peptides Y669 and Y670 (residues 706-715 and 716-725, respectively, of the 20kDa domain). Further experiments are required to see whether the effect of peptide Y670 at high concentrations is significant and peptide specific.

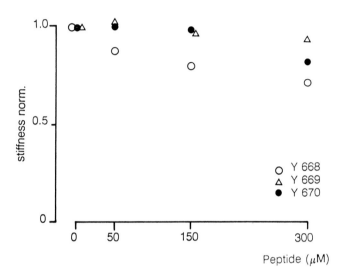

Fig. 1 Effect of peptides Y668, Y669, and Y670 on relaxed fiber stiffness. Speed of stretch 10^4-10^5(nm/half-sarcomere)s^{-1}. Ionic strength = 50mM.

Equatorial X-ray diffraction studies showed equivalent results. However, at high concentrations, all peptides caused broadening and a decrease of the equatorial reflections indicating changes in the submicroscopic structure, but the concentration at which such effects start to occur differs for the various peptides (e.g. about 0.5mM for Y669 but >>1mM for Y668). This effect may result from non-specific binding of the peptides in the fibers.

EFFECT OF MYOSIN PEPTIDES UPON STRONG CROSS-BRIDGE BINDING TO ACTIN

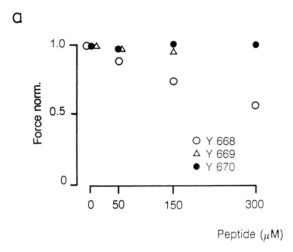

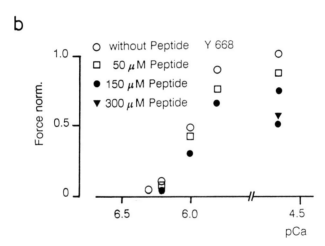

Fig. 2 Effect of peptides on isometric force.
(a) Effect of peptides on maximum isometric force (pCa about 4.5). Ionic strength 170mM.
(b) Changes in force-pCa relation upon addition of various amounts of peptide Y668. Ionic strength 170mM.

Interference of the peptides with strong cross-bridge attachment to actin was tested by recording the developed isometric force at maximum Ca^{++}-activation in the presence and in the absence of the peptide. These measurements were complemented by recording force-pCa relations to test whether peptides can also affect the activation level of the contractile system.

Fig. 2a shows the effect of myosin peptides Y668, Y669, and Y670 on maximum isometric force (pCa about 4.5). Inhibition of active force by peptide Y668 occurred within seconds after addition of the peptide to the muscle bath. Furthermore, isometric force returned to its initial value, again within seconds, upon removal of the peptide. The force-pCa relations (Fig. 2b) demonstrate that the reduction in maximum isometric force by peptide Y668 is not the result of a reduced Ca^{++}-sensitivity of the contractile system. Instead, isometric force is inhibited to the same extent at all Ca^{++}-concentrations. Peptides Y669 and Y670 (data not shown) have no detectable effect on active force at any Ca^{++}-concentration.

Different from active force, neither peptide Y668 nor the other peptides studied so far had any direct effect on fiber stiffness in the absence of nucleotide (rigor) or in the presence of MgPyrophosphate. Under both conditions all cross-bridges are thought to occupy strong-binding states.

EFFECT OF MYOSIN PEPTIDES UPON FORCE REDEVELOPMENT

Fig. 3 shows the effect of the myosin peptide Y668 on the rate constant of force redevelopment (k_{redev}). Different from reduction of isometric force by lowering free Ca^{++}-concentration, which is accompanied by a reduction in k_{redev} (dashed line in Fig. 3), inhibition of maximum isometric force by the myosin peptide Y668 is not associated with any significant change in the rate constant of force redevelopment.

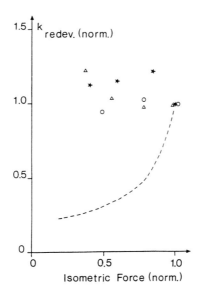

Fig. 3 Effect of increasing concentrations of peptide Y668 on the rate constant of force redevelopment (k_{redev}) at full Ca^{++}-activation (pCa about 4.5). For all concentrations of Y668, k_{redev} is plotted against the observed isometric force. The dashed line represents the decrease in k_{redev} when active force is reduced by lowering free Ca^{++}-concentration. Note that reduction of isometric force by addition of Y668 is not accompanied by a significant change in k_{redev}.

To probe whether the effect of peptide Y668 is sequence specific, a "scrambled" version of this peptide (same amino acids but different sequence: V·R·I·G·R·I·E·L·T·G) was tested. Neither relaxed fiber stiffness nor active force are significantly affected over the concentration range shown in Figs. 1 and 2.

The present study shows that
(i) Of the peptides studied so far only peptide Y668 can interfere with weak cross-bridge binding to actin at µM concentrations.
(ii) The three myosin peptides reported here apparently can not

directly interfere with strong cross-bridge binding to actin (no effect on rigor fibers or in the presence of MgPyrophosphate). However, upon addition of peptide Y668 active force decreases in parallel with inhibition of weak cross-bridge interaction with actin (decrease in relaxed fibers stiffness).

(iii)   While active force is decreased by addition of peptide Y668, cross-bridge turnover kinetics, as judged by the time course of force redevelopment, are unaffected.

(iv)   The effectiveness of peptide Y668 appears to be sequence specific since the scrambled version has no effect on relaxed fiber stiffness or active force.

This situation is identical with the effect of caldesmon or of its actin binding fragments in skinned fibers, suggesting that inhibition of active force by peptide Y668 is not caused by direct inhibition of strong cross-bridge interaction but rather by inhibition of weak cross-bridge attachment to actin which is an essential step in the pathway to force generation (Brenner et al., 1990, 1991; Kraft et al., 1991).

Cross-linking   (Mornet   et   al.,   1981;   Sutoh,   1983), experiments with isolated domains of the myosin head (Chaussepied et al., 1986; Muhlrad, et al., 1986), ^{1}H-NMR studies (Griffiths et al., 1987), and biochemical studies with small peptides (Suzuki et al., 1987) suggested that the region around the SH_1 thiol group (on residue 705) is involved in binding of myosin subfragment 1 to actin.   Keane et al. (1990) demonstrated that the peptide Y668 (residues 697-706) binds to actin with a 1:1 stoichiometry.   On this basis, the specific inhibition of weak cross-bridge attachment to actin, seen in the present study, may indicate that residues 697-706 of the 20kDa domain are a specific part of the binding site on the myosin head which is responsible for weak cross-bridge attachment to actin.   The lack of direct inhibition of strong cross-bridge interaction with actin, however, does not necessarily indicate that residues 697-706, if involved in cross-bridge binding to actin, take part in the weak interaction only.   Instead, the binding or the peptide may simply be too weak to effectively compete with strong cross-bridge interaction at the μM concentrations.

RELATION TO PREVIOUS STUDIES

Chase and Kushmerick (1989) reported that a myosin peptide around the SH-1 group (Suzuki et al., 1987), which has a different amino acid sequence than peptide Y668, can affect force generation in skinned muscle fibers. Later Chase et al. (1990) reported that their originally observed effects were due to contaminants which were introduced during the synthesis procedure. After purification, in skinned fibers no specific effects were detectable any more.

All of our peptides were purified according to Keane et al. (1990) and purity was checked by analytical HPLC and by amino acid analysis. Only peptide Y668 showed effects on relaxed fiber stiffness and active force. This makes it unlikely that the effects seen with peptide Y668 are due to a contaminant from the synthesis procedure or due to presumably unspecific binding effects described by Chase et al (1991) for all purified peptides at millimolar concentrations.

CONCLUSIONS

None of the myosin-derived peptides studied by us so far has direct effects on strong cross-bridge binding to actin. However, the myosin peptide Y668 (Keane et al., 1990) appears to selectively interfere with the weak cross-bridge binding to actin, resulting in a parallel decrease in relaxed fiber stiffness and in active force. As indicated by the control with the scrambled version, the effect of peptide Y668 appears to be sequence specific. However, it is as yet unclear whether peptide Y668 exclusively inhibits weak cross-bridge interaction with actin. The lack of inhibition of strong cross-bridge binding to actin may simply result from too weak binding of the peptide to effectively compete with strong cross-bridge attachment in the fibers at the μM concentrations used in this study. Additional experiments are required to clarify this point and to examine other parts of the myosin head for their involvement in weak and strong cross-bridge interaction with actin. However, the results presented here may suffice to establish the value of using peptides to probe actin-myosin interactions.

REFERENCES

Brenner B (1980) Effect of free sarcoplasmic Ca^{2+} concentration on maximum unloaded shortening velocity: measurements on single glycerinated rabbit psoas muscle fibres. J Muscle Res Cell Motil 1:409-428

Brenner B (1983) Technique for stabilizing the striation pattern in maximally calcium-activated skinned rabbit psoas fibers. Biophys J 41:99-102

Brenner B (1985) Correlation between the cross-bridge cycle in muscle and the actomyosin ATPase cycle in solution. J Muscle Res Cell Motil 6:659-664

Brenner B, Chalovich JM, Greene LE, Eisenberg E and Schoenberg M (1986) Stiffness of skinned rabbit psoas fibers in MgATP and $MgPP_i$ solution. Biophys J 50:685-691

Brenner B and Eisenberg E (1986) The rate of force generation in muscle: correlation with actomyosin ATPase in solution. Proc Natl Acad Sci USA 83:3542-3546

Brenner B, Schoenberg M, Chalovich JM, Greene, LE and Eisenberg E (1982) Evidence for cross-bridge attachment in relaxed muscle at low ionic strength. Proc Natl Acad Sci USA 79:7288-7291

Brenner B, Yu LC and Podolsky RJ (1984) X-ray diffraction evidence for cross-bridge formation in relaxed muscle fibers at various ionic strengths. Biophys J 46:299-306

Brenner B, Yu LC and Chalovich JM (1990) Caldesmon inhibits generation of active tension in skinned rabbit psoas fibers by blocking attachment of weak binding cross-bridges to actin. Biophys J 57:397a

Brenner B, Yu LC and Chalovich JM (1991) Parallel inhibition of active force and relaxed fiber stiffness in skeletal muscle by caldesmon. Implications for the pathway to force generation. Proc. Natl. Acad. Sci. USA (in press)

Chase PB, Bursell J and Kushmerick MJ (1990) Peptide inhibition of rabbit psoas muscle skinned fiber mechanics & myofibrillar ATPase activity. Biophys J 57:538a

Chase PB and Kushmerick MJ (1989) Ca-dependence of a myosin heavy chain (HC) peptide binding to and dissociation from skinned fibers from the rabbit psoas muscle. Biophys J 55:406a

Chaussepied P, Mornet D, Audemard E and Kassab R (1986) Properties of the alkali light-chain-20-kilodalton fragment complex from skeletal myosin heads. Biochemistry 25:4540-4547

Craig R, Greene LE and Eisenberg E (1985) Structure of the actin-myosin complex in the presence of ATP. Proc Natl Acad Sci USA 82:3247-3251

Eisenberg E and Greene LE (1980) The relation of muscle biochemistry to muscle physiology. Ann Rev Physiol 42:293-309

Eisenberg E and Hill TL (1985) Muscular contraction and free energy transduction in biological systems. Science NY, 227:999-1006

Griffiths AJ, Levine BA and Trayer JP (1987) The interaction of the SH_1/SH_2 region of the myosin heavy chain with actin. Biochem Soc Trans 15:847-875

Hibberd MG and Trentham DR (1986) Relationship between chemical and mechanical events during muscular contraction. Ann Rev Biophys Biophys Chem 15:119-161

Huxley AF (1957) Muscle structure and theories of contraction. Prog Biophys 7:255-318

Huxley AF (1974) Muscular contraction. J Physiol (Lond) 243:1-43

Huxley HE (1969) The mechanism of muscular contraction. Science 164:1356-1366

Keane AM, Trayer IP, Levine BA, Zeugner C and Rüegg JC (1990) Peptide mimetics of an actin-binding site on myosin span two functional domains on actin. Nature 344:265-268

Kraft Th, Chalovich CM, Yu LC and Brenner B (1991) Weak cross-bridge binding is essential for force generation. Evidence at near physiological conditions. Biophys. J. 59 (abstr.)

Mornet D, Bertrand R, Pantel P, Audemard E and Kassab R (1981) Structure of the actin-myosin interface. Nature 292:301-306

Muhlrad A, Kasprzak AA, Ue K, Ajtai K and Burghardt TP (1986) Characterization of the isolated 20 kDa and 50 kDa fragments of the myosin head. Biochim Biocphys Acta 869:128-140

Pringle JWS (1967) The contractile mechanism of insect fibrillar muscle. Prog. Biophys. 17:1-60

Rosenfeld SS and Taylor EW (1984) The ATPase mechanism of skeletal and smooth muscle acto-subfragment 1. J Biol Chem 259:11908-11919

Simmons RM and Jewell BR (1974) Mechanics and models of muscular contraction. Recent Adv Physiol 31:87-147

Stein LA, Schwarz RP, Chock PB and Eisenberg E (1979) Mechanism of actomyosin adenosine triphosphate. Evidence that adenosine 5'-triphosphate hydrolysis can occur without dissociation of the actomyosin complex. Biochemistry 18:3895-3909

Sutoh K (1983) Mapping of actin-binding sites on the heavy chain of myosin subfragment 1. Biochemistry 22:1579-1585

Suzuki R, Nishi N, Tokura S and Morita F (1987) F-actin-binding synthetic heptapeptide having the amino acid sequence around the SH1 cysteinyl residue of myosin. J Biol Chem 262:11410-11412

Yu LC and Brenner B (1989) Structures of actomyosin crossbridges in relaxed and rigor muscle fibers. Biophys. J 55:441-453

Yu LC, Maeda Y and Brenner B (1990) Evidence for at least three distinct configurations of attached crossbridges in skinned muscle fibers. Biophys J 57:409a

Caldesmon Derived Polypeptides as Probes of Force Production in Skeletal Muscle

Joseph M. Chalovich[*], Leepo C. Yu[@], Laly Velaz[*], Terry Kraft[#] and Bernhard Brenner[#]
[*]East Carolina University School of Medicine, Greenville, NC, USA
[@]National Institutes of Health, Bethesda, MD, USA
[#]Universitat Ulm, Ulm, FRG

In the work described here, a proteolytic fragment of the smooth muscle protein, caldesmon, is shown to specifically inhibit the binding of myosin subfragment-ATP complexes to actin in solution. This same polypeptide is shown to inhibit the association of weak binding myosin cross-bridges to actin in single skinned rabbit psoas muscle fibers. Inhibition of the weak binding of myosin to actin has several consequences. In solution , the ATPase activity of acto-S-1 or acto-HMM is inhibited, in fibers, active force production is inhibited. These results demonstrate the importance of the weak binding interaction.

Evidence for two types of attached cross-bridges.

The concept of weak and strong binding cross-bridge states evolved from observations of differences in the binding of myosin to actin in the presence of different nucleotides. The weak binding observed in the presence of ATP differs from the strong binding in the absence of ATP by having a lower affinity for actin under identical conditions (Stein et al., 1979) by attaching and detaching to actin more quickly (Stein et al., 1979, Brenner et al., 1986) by binding to actin at a different orientation (Reedy et al., 1965; Moore et al., 1970; Craig et al., 1985) and by binding in a less ordered fashion (Svensson &

Thomas, 1986). In the presence of ADP or absence of nucleotides, the binding of skeletal myosin S-1 or HMM to actin-
tropomyosin-troponin is Ca^{++}-dependent and is cooperative
(Greene & Eisenberg, 1988). The weak binding of myosin to
actin is much less Ca^{++}-sensitive (Chalovich et al., 1981-1986;
El-Saleh & Potter, 1985; Wagner, 1984) and has little if any
cooperativity (Chalovich et al., 1983). This difference, in
response to the regulatory proteins, is maintained even under
conditions where the binding constant of the weak and strong
states are equal (Chalovich et al., 1983). Weak attachment of
myosin cross-bridges to actin has been observed directly in
relaxed muscle fibers (Brenner et al., 1982). In these experiments, rapid stretches were imposed on relaxed, chemically
skinned single psoas fibers. The stiffness (slope of the curve
of force versus displacement) of the relaxed fibers was greater
than 33% of the rigor stiffness indicating a substantial degree
of cross-bridge attachment in relaxed muscle.

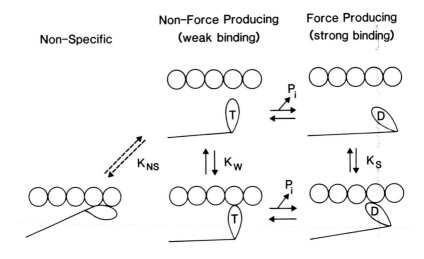

Figure 1. Schematic representation of the cross-bridge cycle
where T represents bound ATP or ADP and Pi and D represents
bound ADP. The equilibrium constants, for the binding of
cross-bridges to actin, are K_w for the weak binding state, K_s
for the strong binding state and K_{NS} for a hypothetical
nonspecific interaction between myosin cross-bridges and actin.

A simplified scheme of contraction incorporating the weak and strong binding cross-bridge configurations is shown in Fig. 1. More detailed descriptions of the cycle may be found elsewhere (Stein et al., 1979, 1988; Rosenfeld and Taylor, 1984; Hibberd and Trentham, 1986; Eisenberg and Greene, 1980). In this model, myosin binds weakly to actin when it contains either bound ATP or bound ADP + Pi. The binding of myosin to actin accelerates the release of Pi and the resulting myosin-ADP complex binds tightly to actin. The release of phosphate may involve one or more intermediate states which are not shown. Force is thought to be produced following phosphate release by the strong binding states (Eisenberg & Greene, 1980) presumably by cross-bridge interactions with actin (Huxley, 1957; Pringle, 1967; Huxley, 1969).

Despite the observation of weak binding cross-bridge states in solution and in fibers, there had been no proof that this state is an essential intermediate in the contractile cycle. It has been suggested that the weak binding state is a nonspecific interaction, indicated by K_{NS} on Fig. 1, which is not a part of the contractile cycle. To demonstrate that the weak binding state is an essential intermediate in the cross-bridge cycle (K_W and not K_{NS} in Fig.1) we showed that inhibition of binding in the weak binding conformation with caldesmon inhibits force production.

Characterization of caldesmon and its 20 kDa actin binding fragments.

The smooth muscle protein, caldesmon, has properties necessary for selective inhibition of weak cross-bridge binding. Caldesmon binds to actin and inhibits actin activation of myosin ATPase activity (Sobue et al, 1981). The inhibition of actomyosin ATPase activity is primarily due to the inhibition of binding of myosin-ATP complexes to actin (Chalovich et al., 1987; Hemric et al., 1988; Velaz et al., 1990; Katayama et al., 1989) although there is disagreement over the latter point (Marston, 1988). The same region of actin appears to be involved in binding to both myosin and caldesmon (Adams et al., 1990; Bartegi et al., 1990 Patchell et al., 1989). Caldesmon also competes with the binding of HMM-AMP-PNP (Hemric & Chalovich, 1988), S-1-PPi (Chalovich et al., 1987), and rigor myosin complexes to actin (Chalovich, 1988; Hemric & Chalovich, 1988). The binding constant of caldesmon to actin, about $8 * 10^6$ M^{-1} at 66 mM ionic strength, (Velaz et al., 1989) is similar to K_s (Chalovich et al., 1983) and about 10^3 stronger than K_w (Chalovich & Eisenberg, 1982). Thus, concentrations of caldesmon which totally inhibit weak cross-bridge binding have little or no effect on strong cross-bridge binding. Another promising feature of caldesmon as a probe is that it had been shown to function when added to skeletal (Galazkiewicz et al., 1987) or smooth (Szpacenko et al., 1985) muscle fibers.

The use of caldesmon for fiber studies has the potential for ambiguity because caldesmon binds to myosin as well as to actin (Hemric & Chalovich, 1988; Ikebe & Reardon, 1988; Hemric & Chalovich, 1990). To insure that the effects of caldesmon are specifically due to its interaction with actin we prepared caldesmon fragments which lacked the ability to bind to myosin. A series of actin binding fragments of caldesmon may be prepared by digestion with chymotrypsin (Szpacenko & Dabrowska, 1986; Fujii et al., 1987; Katayama, 1989). We isolated three fragments with molecular weights of about 20 kDa which originate from the same region of the C terminus of caldesmon and are produced from slightly different cleavage sites. These caldesmon fragments do not bind to myosin. Fig. 2 shows that the 20 kDa actin binding fragments bind to actin with a constant stoichiometry of 1 fragment per 2 actin monomers and an association constant of about 7×10^5 M^{-1}. This binding constant is 1/3 to 1/2 as strong as the binding of intact caldesmon. The inset to Fig. 2 shows that these fragments, like intact caldesmon, inhibit both the actin-activated ATP hydrolysis of myosin and the binding of myosin-ATP to actin. The nearly parallel decrease in both ATPase activity and weakly bound myosin subfragment associated with actin indicate that the inhibition of ATPase activity is largely the result of the inhibition of binding of myosin to actin.

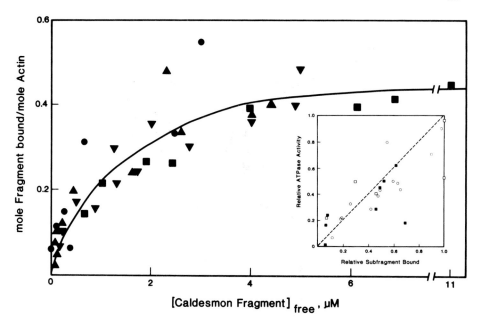

<u>Figure 2.</u> Binding of the 20 kDa caldesmon fragments to actin
and inhibition of actin activated ATPase activity. Binding was
measured in the presence of tropomyosin at 25° C and at 66 mM
ionic strength. The inset shows the correlation between the
inhibition of actin activated ATPase activity of skeletal S-1
(o) and smooth HMM (■,□) and the degree of inhibition of the
binding of S-1 or HMM to actin in the presence of either
caldesmon (o) or caldesmon fragment (□,■).

**The effect of caldesmon and its 20 kDa fragments on muscle
fiber mechanics.**

Our first experiments with skinned psoas fibers were done
with intact caldesmon. Fig. 3 shows that caldesmon enters a
chemically skinned single psoas fiber slowly and reduces the
rapid stiffness of the relaxed fiber. This indicates that the
observed stiffness of the relaxed fiber seen here and earlier
(Brenner et al., 1982; Schoenberg, 1985) is due to the presence
of weak binding cross-bridges. Furthermore because caldesmon

specifically binds to actin and inhibits the binding of myosin to actin in solution, it appears that the weak cross-bridge binding in the fiber is specific also. The inhibition of relaxed stiffness by caldesmon is fully reversible by washing out the caldesmon in a solution containing Ca^{2+}-calmodulin and in which ATP is replaced by PPi to prevent contraction. It is also noteworthy that under conditions where the relaxed stiffness is totally inhibited, there is no effect on rigor stiffness and little, if any, effect on the stiffness in the presence of pyrophosphate.

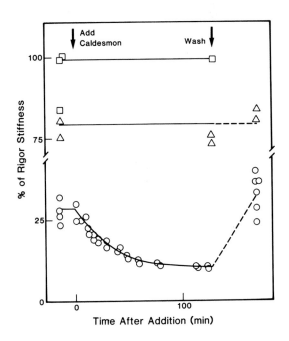

Figure 3. Effect of intact caldesmon on the stiffness of a relaxed rabbit psoas fiber, at 50 mM ionic strength, in the presence of ATP (○), PPi (△), or in rigor (□). The first arrow shows the point at which caldesmon was added to the fiber. At the second arrow, the caldesmon solution was replaced with a 170 mM ionic strength solution containing PPi and calmodulin. The fiber was then placed in the original solution without caldesmon.

Subsequent experiments were done with the 20 kDa actin binding fragments of caldesmon. These fragments enter into single fibers more quickly than intact caldesmon and no further change in fiber mechanics is observed following a 15-20 minute incubation. Like intact caldesmon, the actin binding fragments inhibit relaxed fiber stiffness suggesting that the relaxed stiffness is a measure of specific weak binding cross-bridge attachment. The degree of inhibition of relaxed stiffness is dependent of the added fragment concentration as shown in Fig.4. At very high concentrations of the actin binding fragments (2.6 mg/ml; not shown) the stiffness is reduced to about 25% of the relaxed fiber stiffness in the absence of caldesmon. This residual stiffness is less than 5% of the rigor stiffness. Whether this residual stiffness is due to incomplete access of all actin filaments to the caldesmon fragment or to non-cross-bridge viscoelastic elements of the fiber is not yet known. It is clear form this experiment that the effect of caldesmon results from its interaction with actin and is not dependent on its binding to myosin.

The data shown in Fig. 4. were obtained at $5^{O}C$ and 50 mM ionic strength. The same percentage of inhibition of relaxed fiber stiffness was also obtained at temperatures as high as $20^{O}C$ and at ionic strengths as high as 170 mM. Therefore, the weak binding cross-bridges observed in relaxed fibers are not restricted to low temperature and low ionic strength conditions but are a general property of the contractile system.

Verification that the observed decrease in fiber stiffness results from decreased cross-bridge binding was obtained from X-ray diffraction measurements. It has been previously demonstrated that the intensity ratio of the two innermost reflections, I_{11}/I_{10} decreases as mass is displaced from the thin filament (Huxley, 1968; Yu, 1989). The addition of 0.2 mg/ml actin binding fragment caused an increase in I_{10} by 7% and a decrease in I_{11} by 22% (data not shown). The ratio I_{11}/I_{10} decreased by 27% confirming detachment of myosin cross-bridges.

The specificity of caldesmon for weak cross-bridge states allowed us to determine if the weak cross-bridge states are

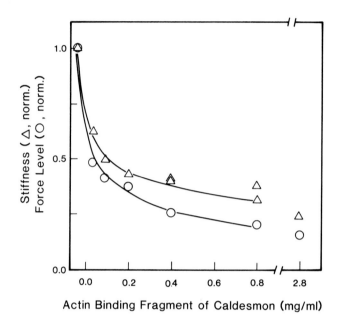

Figure 4. Effect of the 20 kDa actin binding fragments of caldesmon on the stiffness (o) and active tension (Δ) of a relaxed skinned rabbit psoas fiber.

intermediates in the normal cross-bridge cycle. This was done by comparing the relaxed fiber stiffness, at a given caldesmon fragment concentration, with the force that the same fiber is able to produce, upon the addition of Ca^{2+}. If the weak interaction of myosin and actin were not an intermediate in the cross-bridge cycle, like K_{NS}, the addition of caldesmon to a muscle fiber would decrease the population of cross-bridges involved in this nonproductive binding and thereby increase the number of cross-bridges that can participate in force production. In contrast, a decrease in force would occur if the weak binding were a component of the cross-bridge cycle as indicated by K_W in Fig. 1. The results of an actual experiment, done at 50 mM ionic strength, are shown in Fig. 4.

The active force decreased in parallel with the decrease in the weak binding interaction. Under the same conditions, there was no detectable effect on the stiffness, in the presence of PPi or in rigor. Therefore, the weak binding reaction is essential for completion of the cross-bridge cycle. That is, the weak binding observed must represent an interaction, such as K_w, which precedes force production. Similar results were obtained at 120 and 170 mM ionic strength and at 20°C.

In Fig. 4, the active force decreased to a slightly greater extent than the stiffness. Two possible explanations of this behavior are being considered. First, caldesmon may have a slightly greater effect on the binding of weak binding cross-bridges in the presence of Ca^{++} (conditions of force measurements) than in the absence of Ca^{++} (conditions of stiffness measurements). Second, there may be a small degree of inhibition of strong binding cross-bridges which could add to the observed inhibition of force production. These possibilities are under investigation.

To determine if caldesmon could directly alter cross-bridge cycling kinetics we measured the effect of the actin binding fragments on the rate constant of force redevelopment. This measurement had been shown earlier (Brenner, 1988) to provide an estimate of the rate constants of active cross-bridge turnover ($f_{app} + g_{app}$). The rates of attachment and detachment of cross-bridges (k_{+w}, k_{-w}, k_{+s} and k_{-s}) are much faster than the other transitions which occur in the cross-bridge cycle. Thus the free and bound cross-bridges are in rapid equilibrium with respect to the rest of the cycle. The addition of caldesmon results in a reduction of the total force but not in the rate of force redevelopment. Because the rate of force redevelopment is independent of the number of attached cross-bridges, the predominant mechanism of inhibition by caldesmon, in skeletal muscle, is by reduction in the number of bound weak binding cross-bridges. In contrast to the tropomyosin-troponin complex (Brenner, 1988), caldesmon does not effect the rate of force redevelopment.

Discussion

Evidence has been presented, both in solution and in single fibers, that the weak binding conformation of actomyosin is an essential intermediate in the contractile cycle. These results have been obtained at ionic strengths from 50 mM to 170 mM and at temperatures from 5° to 20° showing that this conclusion is not restricted to a particular condition. These results also help to clarify the mechanisms by which caldesmon and tropomyosin-troponin inhibit ATP hydrolysis. We observed earlier that, in solution, caldesmon inhibits actin activated ATP hydrolysis in parallel with inhibition of binding of myosin to actin. The present results show that caldesmon also inhibits weak cross-bridge binding to actin in skeletal muscle fibers. It must be emphasized that caldesmon is not normally found in skeletal muscle and therefore caution must be exercised in extrapolating the skeletal fiber studies to caldesmon function in smooth muscle and non-muscle cells.

The present work also leaves little doubt that tropomyosin-troponin can inhibit actin activated ATPase activity without inhibiting, by a similar extent, the weak binding of myosin to actin. This is made very clear by the striking contrast between the effects of caldesmon and tropomyosin-troponin on the actomyosin interaction in solution and in the fiber. That is, if caldesmon affects cross-bridge binding then tropomyosin-troponin must affect some other process. An idea as to what this other process is can obtained by applying the concept of detailed balance. Because tropomyosin-troponin has a small effect on K_w but a much larger effect on K_s and because tropomyosin-troponin cannot affect the ATP hydrolysis when myosin is detached from actin, some step associated with the transition of AM-ADP-Pi and AM-ADP release must be Ca^{++} sensitive (see Fig. 1 and Chalovich et al., 1981, 1982). These results would be consistent with either a decrease in the forward rate or increase in the reverse rate of one or more processes between the weak and strong attached states. Pre-steady state kinetic studies are consistent with this view (Rosenfeld & Taylor, 1987) although mechanical transients of fibers in response to a rapid Pi jump may indicate that a step which occurs prior to Pi release may be regulated by tropomyosin-troponin (Millar et al., 1990). This may require the introduction of one or more transitions between the attached weak and strong binding states. The introduction of such transitions into the scheme does not alter the conclusion that tropomyosin-troponin inhibits ATP hydrolysis without inhibiting the binding of myosin to actin but by inhibiting the rate of transition between attached weak states and attached strong states. It is noteworthy that it is possible to explain the lag of force behind the change in X-ray diffraction (Kress et al., 1986) by assuming only a 5-fold change in K_W and a large change in the rate of a step associated with the weak to strong transition (Brenner, 1990).

ACKNOWLEDGEMENTS: The authors wish to acknowledge the financial support given by grants from NIH (AR35216 & AR40540-01A1 to JMC), NATO (900257 to BB, LCY and JMC) and DFG (Br849/1-3 to BB).

REFERENCES

Adams S, DasGupta G, Chalovich JM, Reisler E (1990) Immunochemical evidence for the binding of caldesmon to the NH_2-terminal segment of actin. J Biol Chem 265:19652-19657

Bartegi A, Fattoum A, Kassab D (1990) Cross-linking of smooth muscle caldesmon to the NH_2-terminal region of skeletal F-actin. J Biol Chem 265:2231-2237

Brenner B, Schoenberg M, Chalovich JM, Greene LE, Eisenberg E (1982) Evidence for cross-bridge attachment in relaxed muscle at low ionic strength. Proc Natl Acad Sci USA 79:7288-7291

Brenner B, Chalovich JM, Greene LE, Eisenberg E, Schoenberg M (1986) Stiffness of skinned rabbit psoas fibers in MgATP and MgPPi solution. Biophys J 50:685-691

Brenner B (1988) Effect of Ca2+ on cross-bridge turnover kinetics in skinned single rabbit psoas fibers: implications for regulation of muscle contraction. Proc Natl Acad Sci USA 85:3265-3269

Brenner B (1990) Muscle mechanics and biochemical kinetics. In Molecular mechanisms in muscular contraction. J.M. Squire, editor. Macmillan Press, 77-149

Chalovich JM, Chock PB, Eisenberg E (1981) Mechanism of action of troponin-tropomyosin:inhibition of actomyosin ATPase activity without inhibition of myosin binding to actin. J Biol Chem 256:575-578

Chalovich JM, Eisenberg E (1982) Inhibition of actomyosin ATPase activity by troponin-tropomyosin without blocking the binding of myosin to actin. J Biol Chem 257:2432-2437

Chalovich JM, Greene LE, Eisenberg E (1983) Crosslinked myosin subfragment 1: a stable analogue of the subfragment-1·ATP complex. Proc Natl Acad Sci USA 80:4909-4913

Chalovich JM, Eisenberg E (1986) The effect of troponin-tropomyosin on the binding of heavy meromyosin to actin in the presence of ATP. J Biol Chem 261:5088-5093

Chalovich JM, Cornelius P, Benson CE (1987) Caldesmon inhibits skeletal actomyosin subfragment-1 ATPase activity and the binding of myosin subfragment-1 to actin. J Biol Chem 262:5711-5716

Chalovich JM (1988) Caldesmon and thin-filament regulation of muscle contraction. Cell Biophys 12:73-85

Craig R, Greene LE, Eisenberg E (1985) Structure of the actin-myosin complex in the presence of ATP. Proc Natl Acad Sci USA 82:3247-3251.

Eisenberg E, Greene LE (1980) The Relation of Muscle Biochemistry to Muscle Physiology. Ann Rev Physiol 42:293-309.

El-Saleh SC, Potter JD (1985) Calcium-insensitive Binding of Heavy Meromyosin to Regulated Actin at Physiological Ionic Strength. J Biol Chem 260:14775-14779

Fugii T, Imai M, Rosenfeld GC, Bryan J (1987) Domain mapping of chicken gizzard caldesmon. J Biol Chem 262:2757-2763

Galazkiewicz B, Borovikov YS, Dabrowska R (1987) The effect of caldesmon on actin-myosin interaction in skeletal muscle fibers. Biochim Biophys Acta 916:368-368

Greene LE, Eisenberg E (1988) Relationship between regulated actomyosin ATPase activity and cooperative binding of myosin to regulated actin. Cell Biophys 12:59-71

Hemric ME, Chalovich JM (1988) Effect of caldesmon on the ATPase activity and the binding of smooth and skeletal myosin subfragments to actin. J Biol Chem 263:1878-1885

Hemric ME, Chalovich JM (1990) Characterization of Caldesmon Binding to Myosin. J Biol Chem 265:19672-19678

Hibberd MG, Trentham DR (1986) Relationships between chemical and mechanical events during muscular contraction. Annu Rev Biophys Biophys Chem 56:119-161

Millar NC, Lacktis J, Homsher E (1990) The effect of calcium on the phosphate release step of the actomyosin ATPase in skinned psoas muscle fibers. Biophys J 57:397a

Huxley AF (1957) Muscle structure and theory of contraction. Prog Biophys Biophys Chem 7:255-318

Huxley HE (1968) Structural difference between resting and rigor muscle: evidence from intensity changes in the low-angle equatorial X-ray diagram. J Mol Biol 37:507-520

Huxley HE (1969) The mechanism of muscular contraction. Science 164:1356-1366

Kress M, Huxley H, Faruqi AR (1986) Structural changes during activation of frog muscle studied by time-resolved X-ray diffraction. J Mol Biol 188:325-342

Ikebe M, Reardon S (1988) Binding of caldesmon to smooth muscle myosin. J Biol Chem 263:3055-3058

Katayama E (1989) Assignment of the positions of chymotryptic fragments and cysteinyl groups in the primary structure of caldesmon in relation to a conformational change. J Biochem (Tokyo) 106:988-993

Katayama E, Horiuchi KY, Chacko S (1989) Characteristics of the myosin and tropomyosin binding regions of the smooth muscle caldesmon. Biochem Biophys Res Commun 160:1316-1322

Marston S (1988) Aorta caldesmon inhibits actin activation of thiophosphorylated heavy meromyosin Mg^{2+}-ATPase activity by slowing the rate of product release. FEBS Lett 238:147-150

Moore PB, Huxley HE, DeRosier DJ (1970) Three-dimensional reconstruction of F-actin, thin filaments and decorated thin filaments. J Mol Biol 50:279-295

Patchell VB, Perry SV, Moir AJG, Audemard E, Mornet D, Levine BA (1989) Comparative aspects of the molecular basis for thin-filament linked regulation of energy transduction in skeletal and smooth muscle. Biochem Soc Trans 17:901

Pringle JWS (1967) The contractile mechanism of insect fibrillar muscle. Prog Biophys Biophys Chem 17:1-60

Reedy MK, Holmes KC, Tregear RT (1965) Induced changes in orientation of the cross-bridges of glycerinated insect flight muscle. Nature 207:1276-1280

Rosenfeld SS, Taylor EW (1984) The ATPase mechanism of skeletal and smooth muscle acto- subfragment 1. J Biol Chem 259:11908-11919

Rosenfeld SS, Taylor EW (1987) The mechanism of regulation of actomyosin subfragment-1 ATPase. J Biol Chem 262:9984-9993

Schoenberg M (1985) Equilibrium muscle cross-bridge behavior: theoretical considerations. Biophys J 48:467-475

Sobue K, Muramoto Y, Fujita M, Kakiuchi S (1981) Purification of a calmodulin-binding protein from chicken gizzard that interacts with F-actin. Proc Natl Acad Sci USA 78:5652-5655

Stein LA (1988) The modeling of the actomyosin subfragment-1 ATPase activity. Cell Biophysics 12:29-58

Stein LA, Schwarz RP, Jr, Chock PB, Eisenberg E (1979) Mechanism of actomyosin adenosine triphosphatase: evidence that adenosine5'-triphosphate hydrolysis can occur without dissociation of the actomyosin complex. Biochemistry 18:3895-3909

Svensson EC, Thomas DD (1986) ATP induces microsecond rotational motions of myosin heads crosslinked to actin. Biophys J 50:999-1002

Szpacenko A, Dabrowska R (1986) Functional domain of caldesmon FEBS Lett 202:182-186

Szpacenko A, Wagner J, Dabrowska R, Ruegg JC (1985) Caldesmon-induced inhibition of ATPase activity of actomyosin and contraction of skinned fibres of chicken gizzard smooth muscle. FEBS Letters 192:9-12

Velaz L, Hemric ME, Benson CE, Chalovich JM (1989) The binding of caldesmon to actin and its effect on the ATPase activity of soluble myosin subfragments in the presence and absence of tropomyosin. J Biol Chem 264:9602-9610

Velaz L, Ingraham RH, Chalovich JM (1990) Dissociation of the effect of caldesmon on the ATPase activity and on the binding of smooth heavy meromyosin to actin by partial digestion of caldesmon. J Biol Chem 265:2929-2934

Wagner PD (1984) Effect of skeletal myosin light chain 2 on the Ca^{2+}-sensitive interaction of myosin and heavy meromyosin with regulated actin. Biochemistry 23:5950-5956

Yu L.C. (1989) Analysis of equatorial X-ray diffraction patterns from skeletal muscles. Biophys J 55:433-440

Myosin and Troponin Peptides Affect Calcium Sensitivity of Skinned Muscle Fibres

J.Caspar RÜEGG°, Claudia ZEUGNER*, Jennifer E. VAN EYK[+], Robert S. HODGES[+] and Ian P. TRAYER[∞]

°Dept. of Physiology II, University of Heidelberg, [∞]School of Biochemistry, University of Birmingham, [+]Department of Biochemistry, University of Alberta

It is well known that the force developed by the contractile system depends on the calcium ion concentration in the medium surrounding the myofilaments as well as on the Ca^{++}-sensitivity or the calcium responsiveness of the contractile apparatus (Rüegg 1988). We have studied the effect of peptides derived from troponin-I and the 20kDa domain of myosin subfragment-1 which affect the calcium sensitivity of the contractile system. These effects are of particular interest in the case of cardiac muscle, where the relationship between intracellular calcium concentration and force may vary over a wide range. For instance, hearts that are subjected to long periods of ischemia may develop a contracture even at basal levels of free calcium (Allshire et al. 1987, Allen and Orchard 1987). This is probably due to a potentiation of ATPase and force generation occurring at low levels of ATP (Winegrad 1979). Indeed, Güth and Potter (1987) showed, that attached crossbridges increase the calcium responsiveness of skinned fibres at low ATP-concentrations, when some crossbridges are in the nucleotide-free state. Under these conditions, the contractile system is activated, and even potentiated at very low calcium ion concentrations which normally do not elicit a contraction (cf. Weber and Murray

1973). In vitro, such a potentiation also occurs in the presence of excess of myosin subfragment-1, S_1, indicating that there is cooperativity in the binding of S-1 to actin (Greene and Eisenberg 1980). Here we propose as a working hypothesis that it may be the actin-binding region of subfragment-1 around the SH-1 group which by, interacting with actin, is capable of "turning on" actin, so that calcium responsiveness increases. We report here that a synthetic peptide mimic of this region (peptide S-1, residues 687-716, cf. Keane et al. 1990) will increase the calcium responsiveness of skinned cardiac fibres at high, but not at very low concentrations of ATP.

Fibre bundles of right pig ventricle of ~ 200 μm diameter or rabbit psoas fibres were skinned with Triton X-100, attached to an AME-801 force transducer, and relaxed in a well stirred bathing solution (cf. Rüegg et al. 1989) containing (in mM): imidazole, 30; ATP, 10; creatine phosphate, 10; NaN_3, 5; calcium EGTA, 5; $MgCl_2$, 12.5; creatine kinase 380 U ml^{-1}; dithiothreitol, 5; P_i 5. Ionic strength was adjusted to 100 mM with KCl, pH 6.7, 20°C. Contraction was induced by immersion into an analogous solution in which EGTA was replaced by a calcium-EGTA buffer. The buffered calcium ion concentration was determined from the ratio of EGTA to calcium EGTA, essentially according to Fabiato (1979), but using an apparent dissociation constant of 1.6 μM for the calcium EGTA buffer at pH 6.7 and 20°C. ATPase activity of skinned fibres was determined according to Güth and Wojciechowski (1986). Freeze-dried troponin peptides, obtained from R.S. Hodges and myosin peptides, from I.P. Trayer, were dissolved in imidazole buffer adjusted to pH 6.7 before use.

A "calcium sensitizing" myosin peptide

Figure 1 shows that addition of 50 μM peptide SI(687-716) to submaximally calcium-activated skinned cardiac fibres increased contractile force slowly. Note that the time course of peptide action is slower than that produced by altering the free calcium sion concentration. This may be due to the comparatively slow inward diffusion of this thirty residue peptide into the bundle of cardiac ventricle fibres. Alternatively the peptide may also slowly change the filament structure, since peptides of this length may interact with more than one actin monomer (Trayer et al., 1991). Similar effects on the calcium responsiveness are also induced by this peptide in skinned psoas muscle fibres (Fig. 2). Interestingly, the increase in isometric force induced by the peptide is paralleled by a similar increase of calcium activated ATPase-activity in skinned fibres (Table 1).

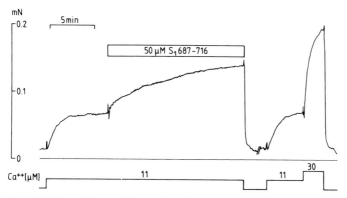

Fig. 1: Effect of a myosin peptide (S_1 687-716) on calcium activated force of skinned pig ventricle fibres. Note reversal of peptide effect after washout (15 minutes).

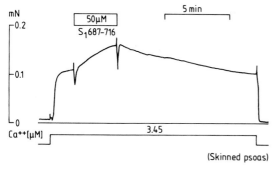

(Skinned psoas)

Fig. 2: Effect of peptide S_1 (687-716) on isometric force of partially activated skinned psoas fibres. "Spikes" on force tracings are solution change artifacts

TABLE 1

Effect of myosin peptide S_1 (689-716) on contractile ATPase and force of skinned cardiac fibres

Conditions	ATPase-activity %	Force %
pCa 5.3 (before addition of peptide)	54	36
pCa 5.3 (with 50 μM peptide)	65	45
pCa 5.3 (after washing out peptide)	51	36

Force and ATPase activity (at 25°C, pH 6,7 after subtracting basal acitivity at pCa 8), given as percentage of maximal activity (at pCa 4.3). Solutions for simultaneous estimation of ATPase and isometric force (Güth and Wojciechowski, 1986) contained (in mM:).

ATP 10	PK 100 U/ml
Imidazole 20	LDH 125 U/ml
$MgCl_2$ 13.7	NADH 6
EGTA; Ca EGTA 5	AP_5 0.2
PEP 5	Pi 5
NaN_3 5	DTT 5

TABLE 2

Effect of myosin peptides on force of skinned cardiac fibres

Peptides (50μM)	Relative force (%)
S_1 (689-716)	154±8 (4)
S_1 (678-687)	103±3 (7)
S_1 (701-728)	108±3 (7)
S_1 (716-725)	102±4 (5)

Control (100%) = force at 8μM Ca^{++} in absence of peptide; numbers in brackets () represent the number of fibre bundles studied.

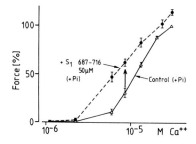

Fig. 3: Effect of peptide S₁ 687-716 (50μM) on Ca-force relation of skinned pig ventricle in presence of 5mM Pi.

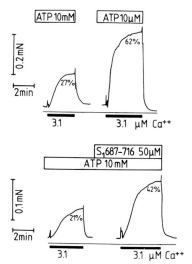

Fig. 4: Effects of ATP concentration and peptide S₁ (687-716) on isometric force (as % of maximal force) of skinned trabecula (pig ventricle).

In both skinned rabbit psoas (Keane et al., 1990) and pig ventricle fibres bundles, addition of the peptide induces a leftward shift of the calcium-force relationship towards higher pCa-values by 0.2 to 0.3 pCa-units (Fig. 3).

Whereas at maximal calcium activation, force was only slightly affected, there is a large potentiation of force production at submaximal calcium activation. This effect mimicks the potentiation effect observed at low ATP-concentrations in skinned cardiac ventricles (Fig. 4).

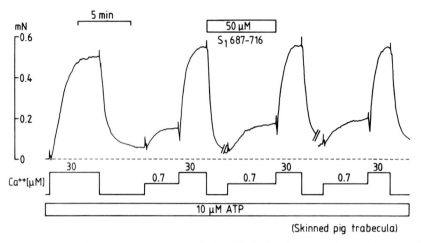

Fig. 5: At low ATP-concentration peptide S_1 687-716 no longer activates submaximal force of skinned pig ventricle fibres

It is noteworthy that at low ATP-concentration, when the calcium responsiveness is increased, addition of the peptide activates contractile force at submaximal calcium activation very little (Fig. 5). Conversely, in skinned fibres, sensitized in presence of the peptide, lowering of the ATP-concentration has a very small additional effect on force.

These experiments may be taken to indicate that the myosin segment containing residues 687-716 is involved in the potentiation effects observed when crossbridges interact with actin at low ATP-concentrations. In support of this notion, we found that peptides derived from segments flanking this region (residues 687-716) had no effect on the calcium responsiveness of skinned cardiac fibres (Table 2).

The ATP-dependence of the S_1-peptide effect is quite puzzling. In this respect, however, it may be of interest that critical residues of the actin binding peptide may be homologous to peptide segments of ATP-binding sites of the 27kDa domain of subfragment-1 and other ATP-binding proteins. If, as proposed by Morita et al.(1991), the dissociation of the 20kDa domain from actin in presence of ATP is due to the competitive binding between ATP and actin at this site, then the interaction of the 20kDa domain and actin should be increased at low ATP-concentration; this, then might cause the potentiation of contraction, even at low calcium ion concentration.

Troponin-I peptides as Ca⁺⁺-desensitizers

Whereas interaction of subfragment-1 with actin increases calcium responsiveness, the interaction of troponin-I with actin decreases it. This inhibitory effect of
troponin-I on acto S-1-ATPase and contractile force may be mimicked by a synthetic
peptide comprised of residues 104-115 of skeletal muscle troponin-I or residues 137-
148 of cardiac troponin-I. These 12 residues are conserved in cardiac and skeletal troponin-I with one exception: the proline at residue 110 of skeletal troponin is replaced by
threonine in cardiac troponin. As shown in Fig. 6, both TnI-peptides have a similar
calcium desensitizing effect on skinned fibres from either skeletal or cardiac muscle.
Note the marked rightward shift (by about 0.2 pCa-units) in the presence of 50 μM
peptide, while there is little decrease in the maximal force. These inhibitory effects are
dose-dependent and half-maximal inhibition occurrs at approximately 30 μM concentration of peptide.

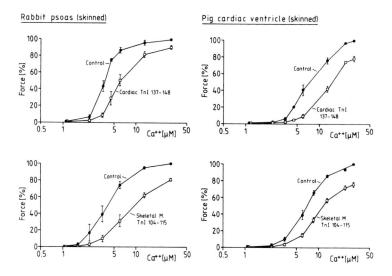

Fig. 6: Effect of TnI-peptides on Ca-sensitivity of skinned fibres

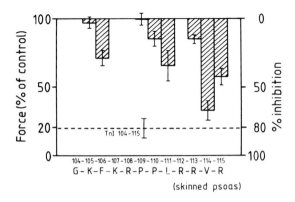

Fig. 7: Effect of different gly-analogs of TnI 104-115 peptide on force of skinned fibres (rabbit psoas). Numbers on abscissa refer to the number of residues that have been replaced by glycine. Submaximal Ca-activation (2.5 or 5 μM Ca^{2+})

Since the TnI peptides are essentially polycations (6 positive charges), it was of interest to determine whether their inhibitory action was merely due to electrostatic effects. Therefore, we tested the effect of different peptide analogs, in which a particular aminoacid residue had been replaced by glycine. Depending on which residue was replaced, the peptide analog produced an inhibitory effect equal to, or smaller than the original native troponin-I peptide, or had no effect at all on skinned skeletal fibres (Fig. 7). In the latter case, the replaced amino acid residue was obviously functionally important, but in the first case it was not. For example, the replacement of the arginine at residue 115 by glycine did not abolish the inhibitory effect of troponin-I peptide suggesting that the high density of positive charges within the peptide was not crucial for its inhibitory action. On the other hand, the replacement of proline 109 abolished the inhibitory activity of the peptide. In general, the effect of amino acid substitutions or "point mutations" of the troponin-I peptide (104-115) was independent of whether skinned skeletal or skinned cardiac muscle fibres were used as a test system. However there was one noteworthy exception: when residue 111 was replaced by glycine, the peptide analog (Gly 111) inhibited skinned skeletal muscle fibres, but surprisingly, activated cardiac muscle fibres (Fig. 8). This peptide analog, therefore, detects molecular differences between the calcium regulatory system of cardiac and skeletal muscle fibres. In interpreting this result, one has to consider that the Troponin-I segment comprising amino acid residues 104-115 interacts in situ not only with actin but also with troponin-C, and these interactions may be affected to different extents by the various peptide analogs depending upon whether skeletal or cardiac muscle fibres are used. These results may be taken to mean that in the skeletal muscle system, the peptide analog inhibits the TnI-TnC-interaction, thereby decreasing calcium sensitivity whereas in cardiac muscle it might predominantly inhibit the TnI-actin interaction rather than the TnI-TnC inter-action, thereby increasing calcium sensitivity. This hypothesis could be tested, if it could be shown that in the skeletal muscle system the peptide has a higher affinity for tropo-nin-C than for actin, whereas in the cardiac contractile system, it would be the other way round.

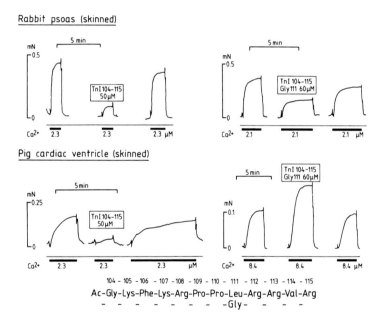

Fig. 8: Effect of TnI (104-115) peptide and its gly 111-analog on force of Ca-activated skinned cardiac and skeletal fibres. Activation of cardiac fibre bundles by the gly 111 analog amounted to 168±7% (n=9) of control!

TABLE 3

Effect of Cardiac TnI (137-148) peptide analogs on skinned cardiac fibres

Peptides (50μM)	Relative force (%) at 6 μM Ca^{++}
TnI 137-148	23±4 (4)
Gly 139	103±4 (6)
Gly 145	107±5 (6)

Control (100%) = force at 6μM Ca^{++} in the absence of peptide. Numbers in brackets () represent number of fibre bundles (right pig ventricle) tested.

Table 3 compares the effects of various analogs of the cardiac troponin-I peptide TnI (137-148) on force production in cardiac fibres. Note that the replacement of various residues with glycine abolishes the inhibitory effect of the peptide analogs on the skinned fibres. The effect of the glycine-145 analog on cardiac skinned fibre is particularly interesting as this peptide has no effect per se, but nevertheless blocks (at 50μM concentration) the inhibitory effect of BDM (2,3 butanedion-2-monoxyme, cf.Horiuti et al., 1988) on submaximally calcium-activated cardiac skinned fibres (Fig.9). Larger concentrations of the drug can overcome this blocking action by the peptide as also shown in Fig. 9.

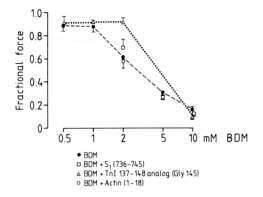

Fig. 9: The dependence of force on concentration of BDM at submaximal calcium activation in presence and absence of peptides in skinned cardiac fibres.

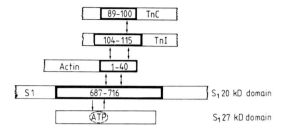

Fig. 10: Interaction of regulatory and contractile proteins in muscle activation; numbers refer to amino acid residues of peptide segments involved. Explanation see text.

Thus, in the presence of this analog, a higher concentration of BDM is required to produce inhibition. This raises the question, whether BDM and the peptide compete for the same binding site. This effect by the Gly 145 analog is quite specific, since other peptides which do not bind to actin, such as myosin S-1 (736-745) or actin (1-18), do not block the BDM-effect. These experiments suggest, that peptide competition experiments may also be used perhaps to identify binding sites of drugs within interacting sequences of regulatory and contractile proteins (cf. Fig. 10).

Discussion

The use of synthetic peptides which mimic functionally important regions of contractile and regulatory proteins, in this and other studies (see Van Eyk et al. 1991, Kraft et al. 1991, Trayer et al. 1991), is a promising approach to increase our understanding of the calcium-dependent protein switches involved in muscle contraction. The model (cf. Fig. 10) that emerges has residues 81-100 of troponin-C interacting more strongly with the inhibitory region of TnI (residues 104-115), when calcium occupies the calcium-specific binding sites (I and II) of troponin-C, whereas the interaction between troponin-I, residues 104-115, and actin become weaker. Interestingly the interaction between troponin-I with troponin-C increases the calcium affinity of troponin-C and hence calcium responsiveness. The interaction between TnI and TnC may be decreased due to peptide competition when using suitable TnI peptides such as TnI (104-115) leading to Ca^{++}-desensitization (Rüegg et al., 1989). The experiments using the peptide TnI-analogs on skinned fibres clearly emphasize the importance of the troponin-I inhibitory region, residues 104-115, of skeletal TnI or residues 137-148 of cardiac muscle TnI, in the regulation of calcium sensitivity. Of course, one might argue that TnI peptides exert their inhibitory action by interacting with actin rather than by peptide competition of TnI-TnC interaction with actin. If this were so, the various TnI peptide analogs should always have a similar effect on acto-S_1 ATPase and force of skinned fibres. However, this is not the case (cf. Fig. 7 and van Eyk and Hodges, 1988).)

In addition to troponin-I, the N-terminal region of actin also acts as a chemical switch (flip-flop mechanism), between a binding site on troponin-I and a binding site on myosin. During calcium activation, the N-terminal segment of actin is bound to the 20kDa-domain of myosin rather than Troponin-I. Indeed, the interaction between actin and myosin activates the myosin ATPase activity.

Actin-myosin interaction and hence acto-S_1 ATPase in vitro may then be inhibited by peptide competition using peptide mimetics of the actin binding site of myosin (Suzuki et al. 1990, Keane et al. 1990) It is therefore puzzling that certain actin-binding peptides e.g. S_1 (687-716) do not inhibit force of skinned fibres at submaximal calcium activation, but enhance it (cf. also Keane et al., 1990). Perhaps these peptides are capable of mimicking the "potentiating" effect (cf. Green and Eisenberg 1980) of nucleotide free Myosin-S_1 interacting with actin. According to Morita et al. (1991), actin activation of myosin ATPase occurs, whenever the inhibitory effect of the 20kDa-domain of myosin on the catalytic site of the acto-S-1-ATPase (on the 27kDa-domain) is repressed as depicted in Figure 10.

In conclusion, skinned fibre studies using synthetic peptides, peptide mimetics, which mimic important regions involved in the interaction between different contractile and regulatory proteins may be useful in delineating the stretch of amino acid residues which comprises the calcium regulatory domains of myosin subfragment-1 and tropo- nin-I. In addition, amino acid-substituted peptide analogs of the natural peptides can also be used to determine the relative importance of individual amino acid residues in the interactions of various protein domains that are critical for the regulation of muscle contraction and calcium responsiveness.

Acknowledgements: We would like to thank Linda Gränz and Elke Höflein for the technical assistance in the preparation of the manuscript and Dr. John Strauss for his critical reading.

References

ALLEN D.G., ORCHARD (1987): Myocardial contractile function during ischemia and hypoxia. Circ. Res. **60**:153-168.

ALLSHIRE .A, PIPER H.M., CUTHBERTSON K.S.R., COBBOLD P.H. (1987): Cytosolic free Ca2+ in single rat heart cells during anoxia and reoxygenation. Biochem. J. **244**:381-385.

FABIATO A., FABIATO F. (1979): Calculator programs for computing the composition of the solutions containing multiple metals and ligands used for experiments in skinned muscle cells. J. Physiol. (Paris) **75**:463-505.

GREENE L.E., EISENBERG E. (1980): Cooperative binding of myosin subfragment-1 to the actin-troponin-tropomyosin complex. Proc. Natl. Acad. Sci. (USA) **77**:2616-2620.

GÜTH K., POTTER J.D. (1987): Effect of rigor and cycling crossbridges on the structure of troponin-C and on the calcium affinity of the calcium specific regulatory sites in skinned rabbit psoas fibers. J. Biol. Chem. **262**:13627-13635.

GÜTH K., WOJCIECHOWSKI R. (1986): Perfusion cuvette for the simultaneous measurement of mechanical, optical and energetic parameters of skinned muscle fibres. Pflügers Arch. **407**:552-557.

HORIUTI K., HIGUCHI H., UMAZUME Y., KONISHI M., OKAUAKI O., KURIHARA S. (1988): Mechanism of action of 2,3-butanedione 2 monoxime on contraction of frog skeletal muscle fibres. J. Muscle Res. Cell. Motil. **9**:156-164.

KEANE A.M., TRAYER I.P., LEVINE B.A., ZEUGNER C., RÜEGG J.C. (1990): Peptide mimetics of an actin-binding site on myosin span two functional domains on actin. Nature **344**:265-268.

KRAFT Th., TRAYER I.P., BRENNER B. (1991): Myosin peptides and cross-bridge kinetics. In Rüegg JC (ed): Peptides as probes in muscle research. Springer Verlag Heidelberg (this volume).

MORITA F., KATOH T., SUZUKI R., ISONISHI K., HORI K., ETO M. (1991): An Actin-Binding Site on Myosin. In Rüegg JC (ed): Peptides as probes in muscle research. Springer Verlag - Heidelberg (this volume).

RÜEGG J.C. (1988): Calcium in Muscle Activation. Springer-Verlag. Berlin-Heidelberg-New York-London-Paris-Tokyo, 2nd printing.

RÜEGG J.C., ZEUGNER C., VAN EYK J., KAY C.M., HODGES R.S. (1989): Inhibition of TnI-TnC interaction and contraction of skinned muscle fibres by the synthetic peptide TnI (104-115) Pflügers Arch. **414**:430-436.

SUZUKI R., MORITA F., NISHI N., TOKURA S. (1990): Inhibition of Actomyos in Subfragment-1 ATPase Activity by Analog Peptides of the Actin-binding Site around the Cys(SH1) of Myosin Heavy Chain. J. Biol. Chem. **265**:4939-4943.

TRAYER I.P., KEANE A.M., MURAD Z., RÜEGG J.C., SMITH K.J. (1991): The use of peptide mimetics to define the actin-binding sites on the head of the myosin molecule In: Rüegg JC (ed): Peptides as probes in muscle research. Springer Verlag - Heidelberg (this volume).

VAN EYK J., HODGES R. (1988): The biological importance of each amino acid residue of the troponin I inhibitory sequence 104-115 in the interaction with troponin C and tropomyosin-actin. J. Biol. Chem. **263**:1726-1732.

VAN EYK J., SÖNNICHSEN F.D., SYKES B.D., HODGES R.S. (1991): Interaction of actin 1-28 with myosin and troponin-I and the importance of these interactions to muscle regulation In: Rüegg JC (ed): Peptides as probes in muscle research. Springer Verlag - Heidelberg (this volume).

WINEGRAD S. (1979): Electromechanical coupling in heart muscle. In: Berne RM, Sperelakis N, Geiger SR (eds) The cardiovascular system. Am. Physiol. Soc. Bethesda, pp 393-428 (Handbook of phisiology, sect 2, vol I).

WEBER A., MURRAY J.M. (1973): Molecular control mechanisms in muscle contraction. Physiol. Rev. **53**:612-673.

Peptides as Probes of the Mechanisms Regulating Smooth Muscle Contractility: Studies on Skinned Fibres

Richard J. Paul, John D. Strauss and Primal de Lanerolle[*].
Departments of Physiology and Biophysics, University of Cincinnati College of Medicine, Cincinnati, Ohio, 45267-0576, USA and [*]University of Illinois, Medical School at Chicago, Chicago, Illinois, 60608, USA.

INTRODUCTION

Our knowledge of the regulatory mechanisms underlying smooth muscle contractility has evolved dramatically over the past 15 years. This growth in knowledge has been paralleled by a wealth of experimental evidence needed to be reconciled with mechanism. To this end, the use of specifically designed peptides in conjunction with the structured and functional permeabilized or skinned fiber preparation appears to be a most promising new technology.

There is a general consensus that phosphorylation of the 20 kDa myosin light chain is a prerequisite for activation of contractile activity (for review, see de Lanerolle and Paul, 1991). Ca^{2+}-sensitivity is conferred by the dependence of myosin light chain kinase (MLCK) on Ca^{2+}/calmodulin. However there is a variety of evidence which suggests that other Ca^{2+}-dependent modulatory systems may be involved. In part, this involves evidence suggesting the independent regulation of isometric force, reflecting crossbridge number, and contraction velocity, an index of crossbridge cycling rate. During an extended contraction, isometric force monotonically increases to a steady maintained value, while shortening velocity decreases from a peak value attained early in the contraction. $MLC-P_i$ also shows a decrease from a maximum value during stimulation, however the extent of this transient is variable, and to a degree, controversial. This state of relatively high force with a decreased velocity of shortening has been termed "latch" (Dillon and Murphy, 1982). Either or both of these parameters have been suggested to be correlated with myosin light chain phosphorylation ($MLC-P_i$). Based on computer modelling (Hai and Murphy, 1989), myosin phosphorylation alone has been suggested to be sufficient to explain the observed relations between force, velocity and $MLC-P_i$, though not necessarily current energetics data (Paul, 1990). Alternatively, other Ca^{2+}-dependent proteins, such as leiotonin (Ebashi et al., 1977), caldesmon (Sobue et al., 1981) and calponin (Winder and Walsh, 1990) have been proposed to play a role in this latch phenomenon. Thus as our data base has increased, our understanding of the mechanisms underlying the Ca^{2+}-sensitivity of smooth muscle have also grown in complexity.

Our investigations have utilized peptides as probes of the role of myosin light chain kinase in regulating contractility utilizing permeabilized fiber preparations. In addition to permitting control over ionic and metabolite concentrations, the permeabilized fibers permit direct access of the peptides to the contractile proteins in a structured, functional, smooth muscle model. Our strategy is to utilize specific peptides to perturb MLCK function at constant levels of Ca^{2+} to unmask other potential Ca^{2+}-dependent regulatory mechanisms.

SKINNED SMOOTH MUSCLE FIBER PREPARATIONS

There are a variety of skinning or permeabilization techniques dependent on the underlying experimental goals and the tolerance of the smooth muscle preparation (Meisheri, Rüegg and Paul, 1985). To achieve uniform access of peptides, our preparation involves exposure to Triton X-100 followed by treatment with a 50% mixture of glycerol and a balanced salt solution. The guinea pig taenia coli is one smooth muscle which yields a very mechanically reproducible and stable preparation following this treatment. This procedure removes or disrupts all membrane systems and this preparation is permeable to large macromolecules such as MLCK (Mrwa et al., 1985), phosphatases (Rüegg, DiSalvo and Paul, 1982) and antibodies (de Lanerolle et al., 1991; see also this volume). It must always be borne in mind that if these macromolecules can access the cell interior, there is the possibility that macromolecules can also exit the fiber. Though this always must be borne in mind, this can in fact be an experimental advantage. For example, in this preparation, its apparent Ca^{2+}-sensitivity is dependent on the level of calmodulin in the bath (Rüegg and Paul, 1982), demonstrating the importance of calmodulin for activation of smooth muscle.

CALMODULIN-BINDING PEPTIDES

As MLCK is dependent on Ca^{2+}-calmodulin for activity, inhibition of kinase in the presence of Ca^{2+} could be achieved by effectors of calmodulin-MLCK interaction. Several agents, such as trifluoperazine (Cassidy, Hoar and Kerrick, 1980) and other putative calmodulin antagonists (Asano, 1989) have been reported to inhibit contraction in smooth muscle. However, the specificity of their actions is open to question (Hartshorne, 1987). Rüegg and colleagues (1989) were the first to use peptides in conjunction with skinned fibers. To achieve a high level of specificity for inhibition at the calmodulin site, they exploited a class of peptides based on a sequence derived from the calmodulin-binding region of myosin light chain kinase developed by Lukas et al. (1986). These peptides bind calmodulin with high affinity (Kd ≡ 10^9) and thus compete with MLCK for available calmodulin.

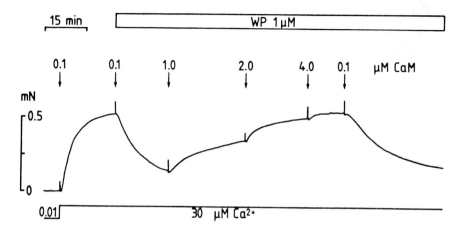

Figure 1. The effects of a calmodulin-binding peptide on isometric force in skinned guinea pig taenia coli. Contraction was elicited by changing the buffered Ca²⁺ concentration from 0.01 to 30 μM in the bathing solution. Note that [Ca²⁺] thereafter remains at 30 μM, and there was initially 0.1 μM calmodulin in the bathing solution. Addition of the calmodulin-binding peptide (1 μM), RS20 or WP as designated in the Figure, immediately elicited a relaxation. Further addition of calmodulin reversed this inhibition, which was reversible as shown by the return of the fiber to lower levels of calmodulin in the presence of WP. For further details, see Rüegg et al., 1989.

The effects of the calmodulin-binding peptide, RS20, on our skinned guinea pig taenia coli preparation are shown in Fig. 1. As expected from the dependence of MLCK on calmodulin, addition of RS20 inhibited both isometric force and MLC-P_i, despite the maintenance of high levels (30 μM) of Ca²⁺ in the bath. Addition of calmodulin in excess of RS20 could reverse this inhibition. Measurements of MLC-P_i paralleled those of force. These experiments demonstrated that a calmodulin-sensitive system underlies the Ca²⁺-regulatory mechanism in this skinned fiber preparation.

Given their high affinities for calmodulin, these peptides can be used as a "calmodulin-buffering" system, similar to that of EGTA for buffering Ca²⁺. These experiments using this peptide as a buffer for calmodulin showed that calmodulin in the nanomolar range was sufficient to activate smooth muscle. This reconciled previous evidence which demonstrated that activity in isolated MLCK was sensitive to calmodulin in the nanomolar range, whereas the apparent sensitivity of skinned fibers for calmodulin was in the μmolar range in the absence of a buffer system. Presumably other cellular calmodulin-binding proteins underlie this difference.

Similar results have been obtained using this class of peptides on inhibition of contractility in skinned single smooth muscle cells (Kargacin, 1990) and in intact single cells, in which RS20 has been pressure injected (Itoh et al., 1989). The latter is particularly noteworthy for it suggests that similar regulatory mechanisms can be demonstrated in intact cells. This would argue against the loss of other potential regulatory systems in the preparation of these skinned fibers.

Studies with the RS20 peptide have demonstrated the absolute dependence of activation of smooth muscle on Ca^{2+}/calmodulin. However, while MLCK would appear to be the major player in the regulation of smooth muscle contractility, other proteins, such as caldesmon and calponin, also dependent on Ca^{2+}/calmodulin cannot be excluded by these data.

MYOSIN KINASE INHIBITING (MKI) PEPTIDE

To inhibit MLCK at a site other than that involving calmodulin, we have used the MKI peptide. This peptide was developed as a substrate level inhibitor for MLCK, and comprises the 11-19 sequence of the myosin light chain (-KKRAARATS-). This sequence contains serine-19, which is the phosphorylation site generally correlated with activation of smooth muscle. It is a substrate level inhibitor with a K_i of approximately 10 μM against MLCK with light chains as substrate (Pearson, Misconi and Kemp, 1986). There is appreciable evidence as to the amino acids and sequence required for inhibition (Hunt et al., 1989). As MLCK has an extremely narrow substrate specificity, this would appear to be an ideal site to target peptide inhibitors of this enzyme.

We were thus quite surprised that our initial studies indicated that even high concentrations of MKI (400 μM) had little effect on either isometric force or MLC-P_i in skinned guinea pig taenia (Strauss et al., 1990). However, and perhaps more noteworthy, as shown in Fig. 2, this peptide was found to inhibit shortening velocity in a dose-dependent fashion.

These results cannot be readily reconciled with any current theory of regulation. Several hypotheses to explain these observations can be put forth. The MKI peptide may directly interact, as it is a light chain analogue, with myosin heavy chains and subsequently inhibit crossbridge cycling rates. MKI carries 4 positive charges and these results and/or this interaction may involve electrostatic interactions. Perhaps most intriguingly, is that MKI may also serve as an inhibitor of phosphoprotein phosphatases. Since it is a substrate for the kinase, it might appear logical that it could also be a phosphatase substrate. However, in general, development of peptide inhibitors of phosphatases (Blumenthal, 1990) has led to the suspicion that more tertiary structure is required for phosphatase inhibition than that for kinases. However, our data suggest that MKI may indeed inhibit phosphatases in the skinned fiber model. Addition to MKI to fibers maintained in solutions of subthreshold Ca^{2+}, induces a slow contraction and concomitant increase in MLC-P_i. MKI also inhibits the rate

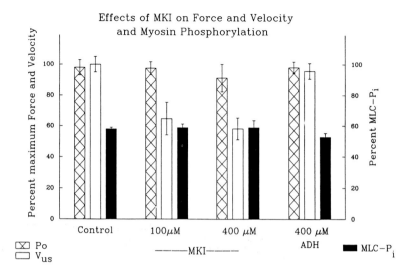

Figure 2. The effects of myosin kinase inhibiting peptide, MKI, on isometric force (Po), velocity (Vus) and myosin light chain phosphorylation (MLC-Pi) in skinned guinea pig taenia coli fibers. Contraction was elicited by a buffered Ca²⁺ concentration of 0.8 μM, which was maintained for all the interventions. ADH indicates the presence of vasopressin, which was used a non-specific peptide control for MKI. For details, see Strauss et al., 1991.

of relaxation, consistent with a role in phosphatase inhibition (Strauss, de Lanerolle and Paul, 1991). Thus MKI has produced some surprising results in skinned fibers, which while suggesting caution in interpretation, also points to the uniqueness and power of this approach.

SUMMARY AND FUTURE DIRECTIONS

The use of specifically designed peptides in conjunction with skinned fibers has provided new and unique information on the mechanisms underlying regulation of smooth muscle contractility. They have also called our attention to differences in the interactions between peptides and muscle proteins in structured systems and those occurring with isolated proteins. Much of the current data is at the level of myosin light chain kinase. Interesting approaches here would include using peptides similar to RS20, but shifted into the so-called pseudo-substrate region (Kemp and Stull, 1990). These peptides are reported to be inhibitory, but do not strongly bind calmodulin. Clearly an area of significance will be the development of peptide inhibitors of phosphoprotein phosphatase. Currently little is known about the nature of the physiologically relevant phosphatase(s) underlying the dephosphorylation of myosin light chains. There is substantial evidence that the substrate specificity of phosphatases may be dramatically altered with the degree of purification of the phosphatases. Thus one might anticipate that our knowledge based on in vitro phosphatase specificity may well be altered when the phosphatase is in a structured environment. Peptides in conjunction with permeabilized preparations could likely play an important role in this area, in particular, and in terms of our overall understanding of the regulation of smooth muscle contractility.

ACKNOWLEDGEMENTS

We would like to thank Professor Dr. Rüegg for his many insightful comments in this area. Supported in part by NIH Grants HL35808, HL02411 (PdeL), HL22619, HL23240 (RJP) and HL TG07571, and the Albert Ryan Foundation (JDS).

REFERENCES

Asano M (1989) Divergent pharmacological effects of 3 calmodulin antagonists, n-(6-aminohexyl)-5-chloro-1- naphthalene-sulfonamide (w-7), chlorpromazine and calmidazolium, on isometric tension development and myosin light chain phosphorylation in intact bovine tracheal smooth muscle. J. Pharm. Exp. Ther. 251(2):764-773.

Blumenthal DK (1990) Use of synthetic peptides to study substrate specificity of protein phosphatases. In: **Peptides and Protein Phosphorylation.** Kemp BE, ed., CRC Press, Inc., Boca Raton, Florida. pp 116-133, 1990.

Cassidy PS, Hoar PE, and Kerrick WGL (1980) Inhibition of Ca^{2+}-activated tension and myosin light chain phosphorylation in skinned smooth muscle strips by the phenothiazines. Pflügers Arch. 387:115-120.

de Lanerolle P and Paul RJ (1991) Myosin phosphorylation/dephosphorylation and the regulation of airway smooth muscle contractility. Am. J. Physiol. (in press)

de Lanerolle P, Strauss JD, Felsen R. and Paul RJ (1990) The effects of antibodies to myosin light chain kinase on contractility and myosin phosphorylation in chemically skinned smooth muscle. Circ. Res. 68:457-465.

Dillon PF and Murphy RA (1982) Tonic force maintenance with reduced shortening velocity in arterial smooth muscle. Am. J. Physiol. 242:C102-C108.

Ebashi S, Mikawa T, Hirata M, Toyo-oka T and Nonomura Y (1977) Regulatory proteins of smooth muscle. In: **Excitation-Contraction Coupling in Smooth Muscle,** Casteels T, Godfraind T and Rüegg JC (eds) pp325-334. Elsevier/North Holland Biomedical Press, Amsterdam.

Hai C-M, and Murphy RA (1988) Regulation of shortening velocity by cross-bridge phosphorylation in smooth muscle. Am. J. Physiol. 255:C86-C94.

Hartshorne DJ (1987) Biochemistry of the contractile process in smooth muscle. In: **Physiology of the Gastrointestinal Tract.** Vol 1 Johnson LR, editor-in-chief. (2nd edition) Chapter 13: 423-482.

Hunt JT, Floyd DM, Lee VG, Little DK and Moreland S. (1989) Minimum requirements for inhibition of smooth-muscle myosin light-chain kinase by synthetic peptides. Biochem. J. 257:73-78.

Itoh T, Ikebe M, Kargacin GJ, Hartshorne DJ, Kemp BE and Fay FS (1989). Effects of modulators of myosin light-chain kinase activity in single smooth muscle cells. Nature 338: 164-167.

Kargacin GJ, Ikebe M and Fay FS (1990) Peptide modulators of myosin light chain kinase affect smooth muscle cell contraction. Am. J. Physiol. 259: C315-C324.

Kemp BE and Stull JT. (1990) Myosin light chain kinases. In: **Peptides and Protein Phosphorylation**. Kemp BE, ed., CRC Press, Inc., Boca Raton, Florida. pp 116-133.

Lukas TJ, Burgess WH, Prendergast FG, Lau W, Watterson DM (1986) Calmodulin binding domains: characterization of a phosphorylation and calmodulin binding site from myosin light chain kinase. Biochemistry 25: 1458-1464.

Meisheri KD, Rüegg JC and Paul RJ (1985) Smooth muscle contractility: Studies on skinned fiber preparations. In **Calcium and Smooth Muscle Contractility**, Daniels EE and Grover AK (eds) Human Press, NJ, pp. 191-224.

Mrwa, U, Güth K, Rüegg JC, Paul RJ, Bostrom S, Barsotti R and Hartshorne D (1985) Mechanical and biochemical characterization of the contraction elicited by a calcium-independent myosin light chain kinase in chemically skinned smooth muscle. Experientia 41(8): 1002-1005.

Paul, RJ (1990) Smooth muscle energetics and theories of crossbridge regulation. Am. J. Physiol. 258: C369-C375.

Pearson RB, Misconi LY, and Kemp BE. (1986) Smooth muscle myosin kinase requires residues on the COOH-terminal side of the phosphorylation site. J. Biol. Chem. 261: 25-27.

Rüegg, JC and Paul, RJ (1982) Vascular smooth muscle: Calmodulin and cyclic AMP dependent protein kinase alter calcium sensitivity in porcine carotid skinned fibres. Circ. Res. 50: 394-399.

Rüegg JC., DiSalvo J and Paul RJ (1982) Vascular smooth muscle: Myosin phosphatase fractions enhance relaxation. Biochem. Biophys. Res. Commun. 106; 1126-1133.

Rüegg JC, Zeugner C, Strauss JD, Paul RJ, Kemp BE, Chem M, Li A-Y and Hartshorne DJ (1989) A calmodulin-binding peptide relaxes skinned muscle from guinea-pig taenia coli. Pflügers Archiv.-Eur. J. Physiol. 414: 282-285.

Sobue K, Muramoto Y, Fujita M and Kakiuchi S (1981) Purification of a calmodulin-binding protein from chicken gizzard that interacts with F-actin. Proc Natl. Acad Sci USA 78: 5652-5655.

Strauss JD, de Lanerolle P and Paul RJ (1991) The effects of myosin kinase inhibiting peptide on contractility and myosin phosphorylation in chemically skinned smooth muscle (Submitted).

Strauss JD, Szymanski PT, DiSalvo J and Paul RJ (1990) Effects of modulators of myosin phosphorylation on isometric force and shortening velocity in skinned smooth muscle. In **Frontiers in Smooth Muscle Research**, [Progress in Clinical and Biological Research: 327] Sperelakis N and Woods JD (eds), A.R. Liss Publisher, NY, pp. 617-622.

Winder SJ and Walsh MP (1990) Smooth muscle calponin. J. Biol. Chem. 265:10148-10155.

Antibodies as Probes of the Mechanisms Regulating Smooth Muscle Contractility: Studies on Skinned Fibres

Primal de Lanerolle[1], John D. Strauss[2] and Richard J. Paul[2], Departments of Physiology and Biophysics, [1]University of Illinois at Chicago, Chicago, IL and [2]University of Cincinnati, Cincinnati, OH, USA

The techniques for permeabilizing (i.e. "skinning") cells were pioneered by studies on muscle fibers (Filo, Bohr and Rüegg, 1965) and have now been extended to many other cell types. The use of cells with permeabilized membranes has some unique advantages with respect to studying the regulation of cellular processes. The most important of these is the ability to manipulate the internal environment of the cell with a level of sensitivity that is not possible in cells with intact membranes (Knight and Scrutton, 1986).

We have used antibodies to myosin light chain kinase (MLCK) in conjunction with skinned fibers to study the regulation of smooth muscle contraction. MLCK is a Ca^{2+}/calmodulin-dependent enzyme that catalyzes the phosphorylation of the 20 kDa light chain (LC_{20}) of smooth muscle myosin. This phosphorylation reaction is generally accepted as the principal mechanism for activating the contractile mechanism in smooth muscles (de Lanerolle and Paul, 1991). Ca^{2+} sensitivity of this mechanism is conferred through the activation of MLCK by the binding of Ca^{2+}/calmodulin to the enzyme. Therefore, we have used antibodies to manipulate MLCK activity while independently manipulating intracellular Ca^{2+} levels and determined the effects of these manipulations on the contractile properties of skinned smooth muscles.

METHODS

All methods and procedures used in the experiments reported in this paper are described in detail in de Lanerolle et al. (1991). Figs. 1-3 and Table 1 are from the same paper and are printed by permission of the authors. Affinity-purified mouse Fab to the Fc region of human IgG, provided by Ralph Hodd, Jackson Immuno-Research Laboratories, Inc., Avondale, PA, were used as a control Fab.

RESULTS

The affinity-purified goat antibodies to MLCK used in these experiments have very similar characteristics to rabbit antibodies that we have described previously (de Lanerolle et al., 1981). Fab fragments prepared from these goat antibodies (MK-Fab) monospecifically interact with MLCK as determined by a Western blot analysis of taenia coli protein separated by SDS-PAGE. This protein has a higher apparent molecular weight (Mr=150,000) than purified turkey gizzard MLCK (Mr=130,000). This is not surprising because MLCKs are known to have heterogeneous molecular weights (Nishikawa et al., 1984). Control Fab do not react with purified turkey gizzard MLCK or with any of the taenia coli proteins when analyzed in the same manner. MK-Fab also inhibit the phosphorylation of LC_{20} by MLCK, whereas control Fab do not.

The effect of MK-Fab on force was studied by preincubating fibers in "relaxing" solution (Ca^{2+}<1 nM) containing MK-Fab before being transferred to contracting solution. Figure 1 demonstrates that fibers incubated with MK-Fab develop less force than in the initial control contraction. The magnitude of this inhibition was dependent upon the length of preincubation; total inhibition occurred with incubation times greater than 90 min. Untreated fibers or fibers treated with control Fab were able to generate approximately 90 percent of initial force after the same length of time in relaxing solution. The observed inhibition would not washout when MK-Fab treated fibers were reincubated for 90 min in normal relaxing solution. This is consistent with the high affinity of these antibodies for MLCK.

The effect of MK-Fab on force maintenance was also determined (Fig. 2). Following a test contraction, the fibers were recontracted by immersion in contracting solution. After the generation of steady state force, the fibers were either maintained in contracting solution or transferred to contracting solution containing BSA, aldolase, control Fab

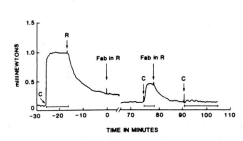

Figure 1: Inhibition of contraction in skinned taenia coli fibers by MK-Fab. C and R refer to contracting (Ca^{2+} = 0.8 μM) and relaxing (Ca^{2+} < 1 nM) solution, respectively, and the horizontal lines indicate the presence of Ca^{2+} in this experiment. Following a test contraction, the fibers were immersed in 4.8 mg/ml MK-Fab (indicated by "Fab in R") dialyzed in relaxing solution. Fibers were recontracted by immersion in high Ca^{2+} solution (indicated by "C") at 75 and 91 mins. The vertical breaks in the force tracing are due to transferring the fiber to a new solution. Force generation was unaffected (compared to control contraction) in other fibers incubated for 120 mins with control Fab dialyzed in relaxing solution.

or MK-Fab.   MK-Fab decreased force despite the continued presence of Ca^{2+}, with a time required for 50% relaxation of about 50 minutes.   Relaxation was rapid when fibers were transferred to relaxing solution at intermediate points during the MK-Fab-mediated relaxation.   In contrast, iso-metric force decreased by about 15% in 4 hours when fibers were incubated with control Fab, BSA or aldolase or when fibers were maintained in contracting solution, alone.   Sur-prisingly, Fab fragments made from protein A purified goat non-immune serum induced complete relaxation within 2 hours.

Because MK-Fab inhibit MLCK activity, it was of inte-rest to determine whether this was the mechanism of action. Therefore, we studied the effects of the MK-Fab and control Fab on fibers previously incubated with ATPγS.   ATPγS as a substrate for MLCK leads to thiophosphorylation of myosin,

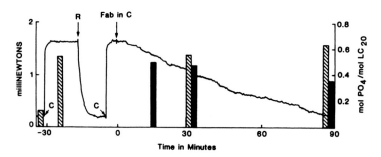

Figure 2: MK-Fab induced inhibition of force maintenance in skinned taenia coli fiber. C and R refer to contracting or relaxing solutions, respectively. At Fab in C, the fibers were incubated with MK-Fab (4.8 mg/ml) dialyzed in contrac-ting solution (0 min). Thus, 0.8 μM Ca^{2+} and MK-Fab were present from 0 to 90 min in this experiment. Other proteins or control Fab had no effect on force maintenance under similar conditions. The columns indicate the myosin phos-phate content in separate control Fab (hatched) and MK-Fab-treated (solid) fibers (n > 5) fixed at the times shown.

which is resistant to dephosphorylation (Cassidy et al, 1979). Therefore, isometric force in thiophosphorylated fibers should be unaffected by inhibiting MLCK catalytic activity. Surprisingly, ATPγS-treated fibers also relax in the presence of MK-Fab.

We measured LC_{20} phosphorylation in skinned fibers to investigate the relationship between myosin dephosphorylation and relaxation. Myosin phosphorylation values are shown in Fig. 2 and are summarized in Table 1. The values obtained with control fibers are similar to those previously obtained from other skinned taenia coli fibers using 2D IEF-SDS PAGE (0.08 and 0.55 mol PO_4/mol LC_{20} in relaxed and contracted fibers, respectively, Hellstrand and Arner, 1985). Our data demonstrate that myosin phosphorylation decreases more slowly than force in fibers treated with MK-Fab. Force is minimal while myosin phosphorylation is still substantially elevated above the resting level at 2 hours (Table 1). Moreover, thiophosphorylated fibers relax completely in the presence of MK-Fab without a statistically significant change in the LC_{20} phosphorylation (Table 1).

Several additional experiments were performed to assess the specificity of these results. The possibility exists that the effect on force maintenance is due to the binding of the MK-Fab to another (i.e. non-MLCK) protein. The most likely candidate for potential cross-reactivity is caldesmond because it is a Ca^{2+}/calmodulin binding protein that can contaminate MLCK preparations (Ngai and Walsh, 1984). To eliminate the possibility that MK-Fab cross-reactivity with caldesmon was responsible for the observed relaxation, we preincubated the MK-Fab for 1 hour with an equimolar concentration of caldesmon (provided by Dr. Michael Walsh,

Univ. of Calgary), before incubation with skinned fibers, to adsorb any Fab which cross-react with caldesmon. The resulting solution relaxed Ca^{2+}-induced contractions with a magnitude and time course similar to those induced by the unadsorbed antibody (Fig. 3A). In contrast, preincubation with an equimolar concentration of MLCK completely eliminated the effect of MK-Fab on force (Fig. 3B). Fibers in Fig. 3B maintained force for over 2 hours and relaxed normally after transfer to Ca^{2+}-free solution.

Table 1

Myosin Phosphorylation and Force in Skinned Fibers

	Isometric Force		mol PO_4/mol LC_{20}
Uncontracted			
Untreated	0.00 (4)		0.13 ± 0.03
Treated			
Control Fab	0.00 (4)		0.13 ± 0.09
Anti-MLCK Fab	0.06 (6)		0.15 ± 0.03
Contracted			
Untreated	0.85 ± 0.05	(9)	0.62 ± 0.01
Treated			
Control Fab	0.88 ± 0.08	(5)	0.63 ± 0.07
Anti-MLCK Fab	0.10 ± 0.05*	(7)	0.36 ± 0.01*
Thiophosphorylated and Contracted			
Untreated	0.67 ± 0.06	(4)	0.65 ± 0.02
Treated			
Anti-MLCK Fab	0.05 ± 0.01*	(4)	0.59 ± 0.02

Values are means $\pm$ SE measured after a 2 hour incubation under the described conditions. Isometric force is given as a fraction of the force generated by the test contraction (see Figs. 1 & 2). The numbers in the parentheses indicate the number of fibers in each group. Control Fab were used at 2.2 mg/ml and MK-Fab were used at 4.8 mg/ml or 0.7 mg/ml. *$p < 0.01$ using a paired t-test when compared to the values for untreated control in each group.

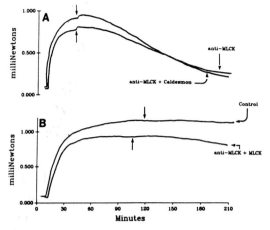

Figure 3: Effect of antibody mixtures on isometric force. Skinned fibers were incubated in contracting solution to achieve a maximal contraction. At the time marked by the unlabeled arrows, the tissues were transferred to baths with fresh contracting solution containing: Panel A: a mixture of 0.85 mg/ml MK-Fab (anti-MLCK) and 2.55 mg/ml caldesmon (anti-MLCK + caldesmon) or 0.85 mg/ml MK-Fab, alone (anti-MLCK); Panel B: 1 mg/ml bovine serum albumin (control) or a mixture of 0.85 mg/ml MK-Fab (anti-MLCK + MLCK); The mixture of Fab and either caldesmon or MLCK was incubated for 1 hour to allow complete interaction before being added to the bath solution.

DISCUSSION

Antibodies, in theory, represent extremely powerful probes for studying regulatory mechanisms because of their specificity. Epitope-specific, affinity-purified polyclonal antibodies or affinity purified monoclonal antibodies can be used with the confidence that they are uniquely interacting with a specific antigen. Our experiments demonstrate that affinity-purified Fab fragments to MLCK (MK-Fab) can affect force and LC_{20} phosphorylation in skinned fibers and establish the feasibility of using antibodies in combination with skinned preparations to study regulatory mechanisms.

Nevertheless, it became clear during our studies that these experiments need to be performed with care and that the choice of appropriate controls is a major issue. Therefore, we performed numerous control experiments to verify the specificity of the Ab-antigen interaction. In our experiments BSA and aldolase had no effect on force development or force maintenance, indicating that protein, per se, has no effect on the fibers. Moreover, aldolase is thought to bind to microfilaments (Hardin et al., 1986) and this control demonstrates that protein interaction with the microfilament network has no effect on force.

Surprisingly, Fab made from protein A-purified non-immune goat IgG also induced relaxation. The reason for this inhibition is not clear to us. These Fab contained mixed populations of antibodies and, presumably, they inhibited force due to the interaction of one or more of the Fab with some protein involved in the contractile process. Analogous effects of antibody preparations on the ATPase activity of actomyosin isolated from smooth muscle have been reported by Mazander et al. (1986). They concluded that the inhibition was perhaps due to the presence of contaminating actin antibodies. We have no direct evidence as to the presence of anti-actin Fab, but it is clear that at the very least, affinity purified antibodies to a specific protein unlikely to be present in smooth muscles are required to permit interpretation of contractility data.

Given the importance of the antibody control experiments, we have studied in detail the effects of affinity purified mouse Fab to human Fc region (control Fab) on the contractile properties of the skinned fibers. The data in Table 1 demonstrate that control Fab affect force only slightly, with a decrease by about 15 percent in 2 hours. None of these fibers relaxed to 50 percent of maximal isometric force in 4 hours, so that T_{50} (the time required for 50 percent relaxation) in these fibers is much greater than 240 min. In contrast, the T_{50} = 49±6 min in MK-Fab treated fibers. The difference in the T_{50}, as well as the data in the Table, clearly demonstrate that the relaxation induced by the MK-Fab is due to a specific effect of the MK-Fab and not due to a non-specific decay in the viability or integrity of the fibers. These data also underscore the importance of using a well-defined, highly purified antibody to a known protein as a control in these experiments.

The data demonstrate that MK-Fab induce complete relaxation in the presence of partial dephosphorylation of the myosin (Fig 2 and Table 1). This is quite surprising and suggests that the MK-Fab have an effect beyond inhibiting MLCK activity. Relaxation of fibers with thiophosphorylated myosin, in the absence of any change in myosin phosphorylation (Table 1), underscores this idea. A potential explanation for this observation is that the MK-Fab are binding to and inhibiting the function of an unidentified protein. This seems unlikely given the monospecific interactions of MK-Fab with MLCK and that adsorption of MK-Fab with MLCK prevents relaxation. Another, more interesting explanation is that MLCK plays a structural role in smooth muscles. MLCK has been co-localized with myosin in cultured cells (de Lanerolle et al., 1981; Cavadore et al., 1982) and it has been suggested that MLCK contains motifs that confer a structural function to the enzyme (Olson et al., 1990). The binding of MK-Fab may alter the filament configuration and/ or impede the actin-myosin interaction. Clearly, the use of site-specific antibodies that bind to these structural motifs in conjunction with skinned fibers could define this putative structural role of MLCK in smooth muscles.

We have recently adapted a voltage discharge technique so that we can introduce μM levels of antibodies and enzymes into cells with intact membranes (Wilson et al. 1991). Our initial experiments, in which we introduced MK-Ab or a constitutively-active form of MLCK into macrophages, indicate that LC_{20} phosphorylation must be maintained within narrow limits to support chemotaxis (Wilson et al., 1989). Taken together with the skinned fiber experiments described above, these experiments establish the feasibility and demonstrate the efficacy of using antibodies to investigate the mechanisms regulating cellular processes.

Supported, in part, by NIH Grants HL35808, HL02411 and HL22619 to PdeL, and HL23240, HL22619 and HL07571 to RJP.

REFERENCES

Cassidy PS, Hoar PE, Kerrick WGL (1980) Inhibition of Ca^{2+}-activated tension and myosin light chain phosphorylation in skinned smooth muscle strips by the phenothiazines. Pflugers Arch 387:115-120

Cavadore JC, Molla A, Harricane MC, Gabrion J, Benyamin Y, DeMaille JG (1982) Subcellular localization of myosin light chain kinase in skeletal, cardiac, and smooth muscles. Proc Natl Acad Sci USA 79:3475-3479

de Lanerolle P, Paul RJ (1991) Myosin phosphorylation/dephosphorylation and the regulation of airway muscle contractility. Am J Physiol, in press

de Lanerolle P, Strauss JD, Felsen R, Doerman GE, Paul RJ (1991) Effects of antibodies to myosin light chain kinase on contractility and myosin phosphorylation in chemically permeabilized smooth muscle. Circ Res 68:457-465

de Lanerolle P, Adelstein RS, Feramisco JR, Burridge K (1981) Characterization of antibodies to smooth muscle myosin kinase and their use in localizing myosin kinase in nonmuscle cells. Proc Natl Acad Sci USA 78:4738-4742

Filo RS, Bohr DF, Rüegg JC (1965) Glycerinated skeletal and smooth muscle: Calcium and magnesium dependence. Science 147:1581-1583

Hardin C, Paul RJ (1986) Association of glycolytic enzymes with chemically skinned guinea pig taenia coli. Fed Proc 45:767

Hellstrand P, Arner A (1985) Myosin light chain phosphorylation and the cross-bridge cycle at low substrate concentration in chemically skinned guinea pig taenia coli. Pflugers Arch 405:323-328

Knight DE, Scrutton MC (1986) Gaining access to the cytosol: The technique and some applications of electropermeabilization. Biochem J 234:497-506.

Mazander KD, Weber M, Groschel-Stewart U, Ngai PK, Walsh MP (1986) The effects of specific antibodies on non-immune globulins on the ATPase activity of smooth and skeletal muscle myosin. Biophys J 49:447a.

Meisheri KD, Rüegg JC, Paul RJ (1985) Studies on skinned fiber preparations. In: *Calcium and Smooth Muscle Contractility*, AK Grover and EE Daniels, eds., Humana Press, New Jersey:191-224

Ngai PK, Walsh MP (1984) Inhibition of smooth muscle actin-activated myosin Mg^{2+}-ATPase activity by caldesmon. J Biol Chem 259:13656-13659

Nishikawa M, de Lanerolle P, Lincoln TM, Adelstein RS (1984) Phosphorylation of mammalian myosin light chain kinases by the catalytic subunit of cyclic AMP- dependent protein kinase and by cyclic GMP-dependent protein kinase. J Biol Chem 259:8429-8436

Olson NJ, Pearson RB, Needleman DS, Hurwitz MY, Kemp BE, Means AR (1990) Regulatory and structural motifs of chicken gizzard myosin light chain kinase. Proc Natl Acad Sci USA 87:2284-2288

Wilson AK, Gorgas G, Claypool WD, de Lanerolle P (1989) An increase or a decrease in myosin II phosphorylation inhibits macrophage motility. J Cell Biology, in press.

Wilson AK, Horwitz J, de Lanerolle P (1991) Evaluation of the electroinjection method for introducing proteins into living cells. Am J Physiol 260:C355-C361.

Regulation of Ca²⁺ Release from Sarcoplasmic Reticulum of Skeletal Muscle by an Endogenous Substance

Annegret Herrmann-Frank
Department of Cell Physiology
Ruhr-Universität Bochum
Universitätsstr. 150
4630 Bochum 1
West Germany

Supported by 'Deutsche Forschungsgemeinschaft' (SFB 114)

INTRODUCTION

Contraction and relaxation of skeletal muscle is controlled by two membrane systems, the transverse tubular and the longitudinal system (sarcoplasmic reticulum). Under physiological conditions, electrical depolarization of the T-tubular membrane triggers a rapid release of Ca^{2+} ions from the sarcoplasmic reticulum (SR), leading to an increase in the intracellular free Ca^{2+} concentration and finally to a contraction of muscle fibre. The communication between the two membrane systems occurs in specialized areas where large protein structures, termed feet or bridges (Franzini-Armstrong, 1986), span the narrow gap between the transverse tubules and the SR. Although the mechanism of signal transmission remains to be elucidated [for recent reviews see Rios & Pizzaro, 1988; Caswell & Brandt, 1989; Fleischer & Inui, 1989], there is now good evidence that SR Ca^{2+} release is mediated via a high conductant ligand-gated Ca^{2+} release channel which serves as the receptor for the plant alkaloid ryanodine and has been shown to be identical with the feet protein [Fleischer & Inui, 1989; Lai & Meissner, 1989; Takeshima et al., 1990]. The SR Ca^{2+} release channel is regulated in a complex manner by Ca^{2+}, ATP and Mg^{2+} and is modulated by various endo- and exogenous ligands including calmodulin, caffeine and the plant alkaloid ryanodine [Lai & Meissner, 1989].

The present chapter describes the effects of an endogenous factor which is released from caffeine-treated amphibian skeletal muscle fibres on Ca^{2+} release from the sarcoplasmic reticulum. Applying the released factor to skinned muscle fibres and isolated 'heavy' SR vesicles, the factor was found to enhance Ca^{2+} release by increasing the mean open time of the Ca^{2+} release channel. Amino acid analysis of the released material suggested that the active component is a small peptide with a molecular mass of about 1800 [Herrmann-Frank & Meissner, 1989].

RESULTS AND DISCUSSION

When skeletal muscles of the frog are immersed in Ringer solution containing low concentration of caffeine, fibres respond with small, irregular oscillations of single sarcomeres that lie below the threshold of contraction [Kumbaraci & Nastuk, 1982; Herrmann-Frank, 1989]. Kumbaraci & Nastuk (1982) first showed that from these oscillating muscle fibres a low-molecular-weight material was released which potentiated caffeine-induced Ca^{2+} release. In the experiments described here, this material has been further purified and its effect has been tested on single skinned skeletal muscle fibres and isolated 'heavy' SR vesicles.

For isolation of the Ca^{2+}-releasing factor, whole muscle preparations of the M. cutaneus pectoris and M. sartorius of *R. tempararia* were used. Typically, eight to ten pairs of muscles were immersed for two hours at $10^{\circ}C$ in Ringer solution containing 2 mM caffeine. Under these conditions, muscle fibres exerted long-lasting, low-frequent sarcomeric oscillations [Herrmann-Frank, 1989]. After two hours of incubation, the bathing solution was collected and evaporated for purification. In a first step, the bathing solution was freed of caffeine by gel permeation on a Sephadex G-10 column. The content of the released material was determined by measuring absorbance at 245 nm. Two major fractions were recovered, one

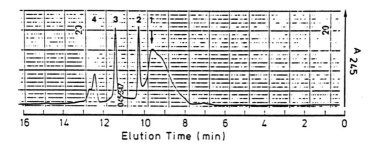

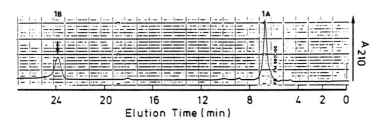

Figure 1: Purification of the Ca^{2+}-releasing factor released from caffeine-treated muscle fibres.

10 pairs of muscles (*M. cutaneus pectoris* & *M. sartorius*) were immersed in 2 mM caffeine Ringer solution for 2 h. The bathing solution was collected, evaporated and freed of caffeine by gel filtration as described by Herrmann-Frank & Meissner (1989). The residual material was lyophilized and loaded on a Polygosil type RP$_{18}$ HPLC column, followed by elution in a buffer medium containing methanol, acetonitrile and water in the ratio 6:10:1 (upper trace). The active material (fraction 1) was further purified by a second elution in a more polar buffer solution containing methanol, acetonitrile and water in the ratio 4:10:1 (lower trace). Only the less polar fraction 1B was found to be active. [From Herrmann-Frank & Meissner (1989)]

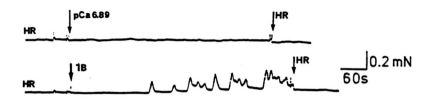

Figure 2: Activity test in skinned skeletal muscle fibres.

A skinned muscle fibre was activated by a slightly EGTA-buffered "Ca^{2+}-releasing solution" [Herrmann-Frank, 1989] containing 60 nM Ca^{2+}_{free}, 0.1 mM Mg^{2+}_{free}, and 0.1 mM EGTA (upper trace). After 3 min in a relaxing solution (HR) containing 50 mM EGTA, the fibre was transferred into a Ca^{2+}-releasing solution of identical pCa additionally containing fraction **1B** of Fig. 1 (lower trace). Oscillations started after short latency of 2-3 min. [From Herrmann-Frank & Meissner (1989)] contained the released material, the other one was related to caffeine. Further purification was achieved in two steps by the means of reversed-phase HPLC. Fractions of gel filtration containing the released material were combined, lyophilized and loaded on a RP-18 HPLC column. Elution in a buffer solution containing methanol, acetonitrile, and water in the ratio 6:10:1 yielded four major fractions (Fig. 1). Testing the effect of each fraction on skinned skeletal muscle fibres, only the first fraction (1) was found to be active. It induced both subthreshold sarcomeric and long-lasting force oscillations, depending on the free Ca^{2+} concentration of the bathing solution. Oscillations ceased when the bathing solution was highly EGTA-buffered or when the membranes of sarcoplasmic reticulum were destroyed by a detergent. The lower part of Fig. 1 shows that the active fraction (1) could be separated into two further components by elution in a more polar buffer medium containing methanol, acetonitrile, and water in the ratio 4:10:1. In this case, only the second eluted component (1B) was active. Fig. 2 demonstrates the activating effect in skinned muscle fibres. To account for small amounts of Ca^{2+} in the active fraction, the fibre was first transferred into a solution with a pCa identical with the test solution containing fraction (1B). While Ca^{2+} by itself did not show an activating effect, the bathing solution containing fraction (1B) caused repetitive contractions. When the free EGTA concentration of the bathing medium was lowered, force oscillations turned into subthreshold sarcomeric oscillations. Again, oscillations were blocked by agents inhibiting Ca^{2+} release from sarcoplasmic reticulum.

Since the active material did not cause any remarkable activation when applied to intact muscle fibres, the results suggested that the component exerted its activating effect by acting on the intracellular SR Ca^{2+} release pathway. To test this hypothesis, two series of experiments were performed: 1- In $^{45}Ca^{2+}$ ion flux measurements, the macroscopic Ca^{2+} release behaviour of heavy SR vesicle was tested in the presence and absence of the released component. 2- Reconstituting

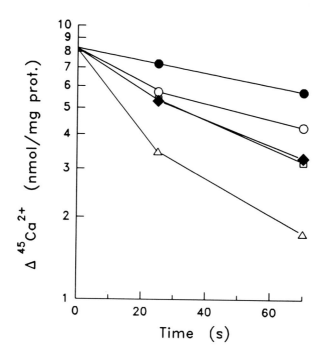

Figure 3: Potentiation of $^{45}Ca^{2+}$ efflux from heavy SR vesicles by the factor released from caffeine-treated muscle fibres.

$^{45}Ca^{2+}$ efflux from 'heavy' skeletal SR vesicles passively loaded with 1 mM $^{45}Ca^{2+}$ was measured by a filtration technique, as described by Meissner [1984]. $^{45}Ca^{2+}$ efflux was initiated by 60-fold dilution into iso-osmolal unlabelled release media [Herrmann-Frank & Meissner, 1989] containing (A): 140 mM Na^+, 20 mM Pipes, 1 mM Mg^{2+}_{free}, 0.1 mM Ca^{2+}_{free} (●), (A) + 3.3 mM AMP (○), (A) + 3.3 mM AMP + 1 mM caffeine (□), (A) + 3.3 mM AMP + 0.5 μg/ml factor **1B** (◆), (A) + 3.3 mM AMP + 3 mM caffeine (△). $^{45}Ca^{2+}$ efflux was stopped by rapid rinsing with a solution containing 5 mM Mg^{2+}_{free} and 10 μM ruthenium red. [From Herrmann-Frank & Meissner (1989), modified] heavy SR vesicles into planar lipid bilayers, the activating effect of the released compound was tested on the single channel level.

Fig. 3 shows a comparison of the effects of low concentrations of caffeine and the active component on SR $^{45}Ca^{2+}$ release. Heavy SR vesicles isolated from rabbit back and leg muscles [Meissner, 1984] were passively loaded with 1 mM $^{45}Ca^{2+}$ for two hours. To reduce the extravesicular free Ca^{2+} concentration to physiological resting conditions, vesicles were initially diluted in a medium containing 1 mM Mg^{2+} and 0.1 μM free Ca^{2+}. After 10 s, a second Ca^{2+} release medium was added containing 0.1 μM free Ca^{2+}, 1 mM free Mg^{2+}, 3.3 mM AMP and the indicated concentrations of caffeine or the released factor, respectively. In the absence of caffeine and the released factor, vesicles released their Ca^{2+} store in about 75 s. An about 2-fold increase in Ca^{2+} efflux rate could be achieved by adding the active component to the release medium. A similar increase in the release rate was obtained in the presence of 1 mM caffeine.

Reconstituting isolated SR vesicles into planar lipid bilayers [Smith et al., 1988] confirmed this result on the single channel level. In Fig. 4, a single Ca^{2+} release channel was initially slightly activated by 0.13 μM Ca^{2+}, displaying a low open probability (P_O) of 0.01. Adding the active fraction to the cytoplasmic side of the channel, the P_O increased 4-fold to about 0.04 (third trace) without any change in the current-voltage relationship. In some cases, immediately after addition of the active compound, single prolonged openings were observed (second trace), turning into the rapid gating of the channel with further incubation time. Increased activity could be inhibited by micromolar concentrations of ruthenium red (fourth trace), indicating that the released factor is acting on the Ca^{2+}-gated Ca^{2+} release pathway of the sarcoplasmic reticulum.

The activating effect on SR Ca^{2+} release raised the question of which class of substances the released factor belonged to. Amino acid analysis suggested that the active component is

138

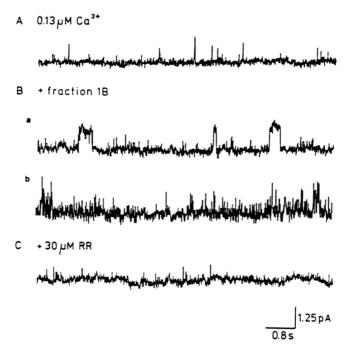

A   0.13 µM Ca²⁺

B   + fraction 1B

a

b

C   + 30 µM RR

1.25 pA

0.8 s

Figure 4:   Activation of a single Ca^{2+} release channel by fraction 1B.

'Heavy' skeletal SR vesicles were fused into Mueller-Rudin planar lipid bilayers containing phosphatidylethanolamine, phosphatidylserine, and phosphatidylcholine in the ratio 5:3:2 (25 mg ml^{-1} phospholipid in decane), as described by Smith et al. (1988). Single channel activity of a subthreshold Ca^{2+}-activated channel was monitored in the absence (A) and presence (B) of fraction 1B. B(a) was recorded immediately after the addition of fraction 1B and B(b) after 8 min of incubation. Increased activity was inhibited by subsequent addition of 30 µM ruthenium red (RR). [From Herrmann-Frank & Meissner (1989)] a small peptide composed of four glycines, two glutamic acids, two serines, two valines, and one threonine, alanine, iso-leucine, asparagic acid and one proline, each, resulting in a molecular weight of 1800 daltons [Herrmann-Frank & Meissner, 1989]. This molecular weight is in the order of magnitude typical of neuropeptides which play a multiple role in the receptor-mediated activation of various tissues [for review see Turner, 1986]. The total amount of amino acids recovered from one purification was 5-6 μg. Between 0.5 μg ml^{-1} (^{45}Ca^{2+} flux and single channel measurements) and 1 μg ml^{-1} (skinned fibres experiments) was needed to cause activity.

CONCLUSIONS

The experiments described here have shown that caffeine-treated skeletal muscle fibres release a small peptide which in turn potentiates Ca^{2+} release from the sarcoplasmic reticulum by increasing the mean open time of the SR Ca^{2+} release channel without affecting its conductance. Since the activating effect is more pronounced when the channel is slightly activated before [Herrmann-Frank & Meissner, 1989], it is suggested that the released compound is capable of enhancing Ca^{2+} release but not of initiating Ca^{2+} release. The fact that the compound is released into the extracellular space raises the question if under physiological conditions the skeletal muscle cell is the only site of action and/or if the factor also modulates physiological processes in other muscle or non-muscle tissues.

REFERENCES

Caswell AH, Brandt NR (1989) Does muscle activation occur by direct mechanical coupling of transverse tubules to sarcoplasmic reticulum? J Bioenerg Biomembr 21:149-162

Fleischer S, Inui M (1989) Biochemistry and biophysics of excitation-contraction coupling. Ann Rev Biophys Chem 18:333-364

Franzini-Armstrong C (1986) The sarcoplasmic reticulum and the transverse tubules. In: Myology (Eds Engel AG & Banker BQ), pp. 125-154. McGraw-Hill New York

Herrmann-Frank, A. (1989) Caffeine- and Ca^{2+}-induced mechanical oscillations in isolated skeletal muscle fibres of the frog. J. Muscle Res. Cell Mot. 10, 437-445

Herrmann-Frank, A. & Meissner, G. (1989) Isolation of a Ca^{2+}-releasing factor from caffeine-treated skeletal muscle fibres and its effect on Ca^{2+} release from sarcoplasmic reticulum. J. Muscle Res. Cell Mot. 10, 427-436

Lai FA, Meissner G (1989) The muscle ryanodine receptor and its intrinsic Ca^{2+} channel activity. J Bioenerg Biomembranes 21: 227-245

Kumbaraci NM, Nastuk WL (1982) Action of caffeine in excitation-contraction coupling of frog skeletal muscle fibres. J. Physiol. 325:195-211

Meissner G (1984) Adenine nucleotide stimulation of Ca^{2+}-induced Ca^{2+} release in sarcoplasmic reticulum. J Biol Chem 259:2365-2374

Rios E, Pizarro G (1988) Voltage sensors and calcium channels of excitation-contraction coupling. NIPS 3:223-227

Smith JS, Coronado R, Meissner G (1988) Techniques for observing calcium channels from skeletal sarcoplasmic reticulum in planar lipid bilayers. Meth Enzymol 157:480-489

Takeshima H, Nishimura S, Matsumoto T, Ishida H, Kangawa K, Minamino N, Matsuo H, Ueda M, Hanaoka M, Hirose T, Numa S (1989) Primary structure and expression from complementary DNA of skeletal muscle ryanodine receptor. Nature 1339: 439-445

Turner AJ (1986) Processing and metabolism of neuropeptides. Essays Biochem 22:70-119

Defining Sites and Mechanisms of Interaction Between Rhodopsin and Transducin

Heidi E. Hamm

Department of Physiology and Biophysics, University of Illinois College of Medicine at Chicago, Box 6998, Chicago, IL 60680, USA.

It has been known for many years that sensory and hormonal effects on cells are mediated by specific receptors which activate diverse intracellular processes. Much recent progress in the understanding of signal transduction has come from molecular and biochemical studies on one large family of signal receptors, the G protein–coupled receptors. These molecules transduce signals by activating GTP–GDP exchange reactions on heterotrimeric G proteins, which in turn interact with diverse effectors to change second messenger levels or ionic gradients in cells. The mechanisms by which receptors interact with and activate GTP binding proteins are currently under intense scrutiny.

Visual signal transduction by photoreceptor cells is initiated by photoisomerization of rhodopsin and subsequent activation of G_t, transducin. Studies of the rod outer segment (ROS) response to light have shown that at least 1000 G protein (G) molecules can be activated by each activated rhodopsin (R^*) (Liebman et al., 1987), and activation of one G protein occurs within a millisecond (Vuong et al., 1984). Within this time, the activated receptor binds to a G protein and two conformational changes occur in the G protein. First, rhodopsin catalyzes the opening of the guanine nucleotide binding pocket of α, releasing GDP, and second, GTP binding triggers a large change leading to loss of affinity for receptor and for $\beta\gamma$ subunit and increased affinity for the effector enzyme, cyclic GMP phosphodiesterase (see Chabre and Deterre for review).

The binding affinity between light-activated rhodopsin and G_t is approximately three orders of magnitude higher than between dark–adapted rhodopsin and G_t (Bennett and Dupont, 1985). The molecular basis for this high–affinity interaction is of great interest. In this report I will concentrate on our recent studies of the determinants of G_t that participate in binding to rhodopsin.

1. Sites of interaction between G_t and rhodopsin.

The regions on rhodopsin and the α_t subunit of G_t involved in binding to rhodopsin were investigated using synthetic peptide competition. Rhodopsin-G_t interaction was measured spectrophotometrically by the stabilization of Metarhodopsin II by G_t binding (Emeis et al., 1982). Three peptides from various parts of the α_t surface competitively blocked R–G interaction (Hamm et al., 1988). Peptides corresponding to the amino and carboxyl termini, as well as an internal peptide corresponding to amino acids 311-329 were able to compete with G_t for binding to rhodopsin. When the two peptides that were the most potent at competing with G_t were added together to the assay, they were 20–fold more potent than either peptide alone. Thus, there is a marked potentiation of the peptide effects when they are presented in combination. It was of interest that one of the synthetic α_t peptides, α_t–340–350, was able to mimic G_t binding to rhodopsin almost completely. In the absence of G_t, α_t–340–350 dose–dependently stabilized metarhodopsin II. All of the data, including the fact that at least two regions on α_t bind rhodopsin, and that they have synergistic effects upon binding, suggests a complex interaction with rhodopsin, including at least two sequences juxtaposed in the tertiary structure of the protein to form a binding site with rhodopsin.

Three loops from rhodopsin's cytoplasmic surface were also able to competitively block R–G_t interaction (Koenig et al., 1990). Combinations of two rhodopsin peptides also had synergistic effects. Thus it appears that there must be a multiple–point interaction between rhodopsin and G_t. To summarize what is known about the rhodopsin–G_t interface, the data on the regions of the two proteins that block the interaction have been incorporated into a model of rhodopsin activation (Fig. 1). In this model, a rather deep G_t binding site on the cytoplasmic region of rhodopsin opens up in the metarhodopsin II conformation. This model points out that in addition to electrostatic interactions between rhodopsin's cytoplasmic loops and the surface–exposed regions of G_t, hydrophobic interaction between the α–helices of rhodopsin and hydrophobic amino acids in α_t is likely to be another stabilizing force in the high–affinity R–G interaction. In fact, a hydrophobic stretch of amino acids connecting the two regions α_t–311–329 and α_t–340–350, α_t–Phe^{332}–His^{340}, is a candidate for such interaction.

2. Determination of the conformation of the rhodopsin binding domain of G_t.

Synthetic peptide α_t–340–350 is the most effective peptide for the direct stabilization of metarhodopsin II. It's effect on stabilization of Meta II is compared with the effect of G_t (Fig. 2a).

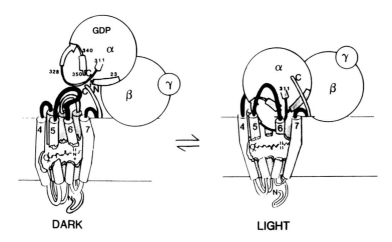

Figure 1. Model of the structural basis of rhodopsin–G_t interaction. The three regions of α_t that block R–G interaction (Hamm et al., 1988) are shown as potential rhodopsin interaction sites. The filled regions on rhodopsin's cytoplasmic surface that block R–G interaction (Koenig et al., 1990) are shown as a complimentary interacting surface.

Its primary structure is given in Fig. 2b. It is of interest that Cys^{347}, four amino acids from the carboxyl terminus of α_t, is the site of ADP–ribosylation by pertussis toxin. G_t as well as at least four other G proteins (G_{i1}, G_{i2}, G_{i3} and G_o) are substrates for pertussis toxin ADP–ribosylation, and it has been shown in many cases that this results in uncoupling of substrate G proteins from their cognate receptors (Kurose et al., 1983; Cote et al., 1984; reviewed in Gilman, 1987). The fact that peptide α_t–340–350 binds to metarhodopsin II and stabilizes it, mimicking the effect of the G_t molecule, strengthens the evidence that this region is a part of the rhodopsin-binding domain of G_t. Additional evidence comes from the fact that another protein that interacts specifically with phosphorylated metarhodopsin II (Wilden et al., 1986), the 48 kDa protein arrestin, shares sequence homology in several regions of α_t including the carboxyl terminal 40 amino acids (Wistow et al., 1986; Hamm et al., 1988).

These results led to an approach for the determination of the conformation of the rhodopsin–binding domain of G_t as it is bound to rhodopsin (Dratz et al., 1990; Dratz et al., 1991; Hamm et al., 1991). Proton nuclear magnetic resonance (NMR) is useful for elucidating molecular conformations in detail, since resonances arise from individual protons in the peptide.

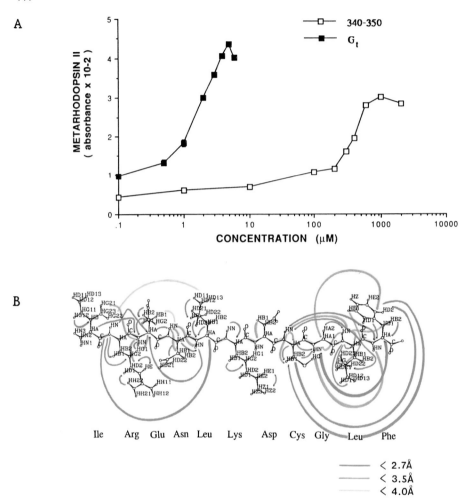

A

B

Ile   Arg   Glu   Asn   Leu   Lys   Asp   Cys   Gly   Leu   Phe

——— < 2.7Å
——— < 3.5Å
——— < 4.0Å

Figure 2. a) Effects of G_t and synthetic peptide α_t–340–350 on Metarhodopsin II (Meta II, Hamm et al., 1988). The filled squares show the amount of Meta II in the presence of varying concentrations of G_t. Synthetic peptide α_t–340–350 dose–dependently mimics G_t and stabilizes Meta II. b) Primary structure of α_t–340–350 showing main short and long range interactions determined by proton NMR (Dratz et al., 1990).

Total correlation spectroscopy (TOCSY) of the peptide in solution shows cross peaks between interacting protons and is used to assign proton resonances to particular amino acids in the peptide. Once the NMR spectrum is assigned, another kind of experiment, rotating frame spectroscopy (ROESY) is used to obtain distances between protons. Protons which are far apart on the primary sequence but close together in solution give an indication of the peptide conformation. When the peptide binds to rhodopsin, its conformation may change. Another type of experiment, transferred nuclear Overhauser effect spectroscopy (trNOESY, Clore and Groenenborn, 1982; Wright et al., 1988; Campbell and Sykes, 1989) can determine the conformation that is enriched in the receptor–bound state. In the presence of metarhodopsin II, the peptide binds and because its affinity is rather low (mM), the peptide on–off exchange rate is rapid. Under these conditions, transferred NOESY peaks are indicative of the bound conformation of the peptide (Dratz et al., 1990, 1991; Hamm et al., 1991). Lines in Fig. 2b connecting various amino acids in the linear sequence show such interactions.

Combining the NMR information with steric constraints and attractive and repulsive forces yielded a model of the conformation of the peptide (Fig. 3). It was predicted to contain a β–turn at its carboxyl terminus and one turn of an imperfect α–helix) at its amino terminus. Since β–turns are known to be key recognition sites in proteins, we were interested in testing whether the β–turn was crucial for G_t binding to rhodopsin. Several other considerations also lent more interest to this region of the peptide. Cys^{347} takes part in the β–turn, and pertussis toxin ADP–ribosylation of this residue uncouples G proteins from receptors. Also, a $Cys^{347} \rightarrow Ser$ substituted analog of α_t–340–350 does not stabilize metarhodopsin II, showing that this cysteine is important for stabilizing some conformation or for binding to rhodopsin.

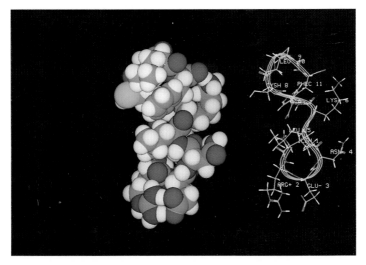

Figure 3. Model of the three–dimensional structure of α_t–340–350 bound to rhodopsin. The trNOESY distances were incorporated using DSPACE distance geometry into a minimum energy conformation. Space–filling (left) and ribbon (right) representations of the model are shown. The peptide appears to have two distinct domains at the carboxyl and amino termini. At the carboxyl terminal of the peptide, Phe^{350} (labelled Phe^{11}) interacts with Cys^{347}, forming a turn composed of Phe^{350}, Leu^{349}, Gly^{348}, and Cys^{347}. The amino terminal five amino acids are interconnected pre-

dominantly through backbone interactions, forming a turn that resembles one turn of an α-helix.

To test the model, molecular dynamics calculations were carried out on the effects of various amino acid substitutions. It was found that Gly[348], a key member of the β-turn, could be substituted with d-alanine and still be predicted to retain the β-turn, while substitution with any other amino acid was predicted to break the turn. Two analogs were synthesized. To stabilize the β-turn (but break other structures), Gly[348] was substituted with d-alanine. To break the β-turn, Gly[348] was substituted with leucine. The d-Ala analog had nearly full potency in stabilizing metarhodopsin II, whereas the Leu analog showed a dramatically decreased potency (Fig. 4). This data is consistent with the importance of a β-turn for recognizing and binding to rhodopsin, and stabilizing the active metarhodopsin II conformation.

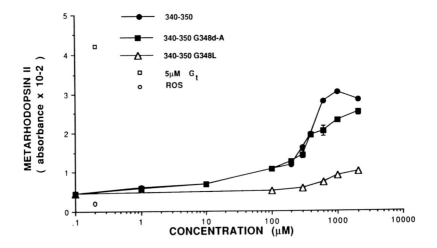

Figure 4. Analog peptides predicted in molecular dynamics simulations to break (L–Leu[348], relative energy 88 kcal/mol) or stabilize (D– Ala[348], relative energy 3.2 kcal/mol) the β-turn were compared to native Gly[348] stabilization of Meta II. The predicted small destabilization of D-Ala[348] was borne out experimentally in the slight decrease in activity of this peptide, while the large destabilization predicted for L–Leu[348] is reflected in a large decrease in its activity.

The amino terminal of the peptide also contains a turn. To test the role of amino terminal residues, truncated peptides from the α_t–340-350 sequence were synthesized. Peptides α_t–345-350 and 344-350 retained less than 20% activity, while peptide α_t–343-350 and 342-350 retained 42 and 27% activity, respectively. This suggests that the amino terminal three amino acids are less important than the carboxyl terminal residues. However, the residual activity of Leu^{348}–340-350 may arise from the amino terminal portion of the peptide.

Comparison of the bound and free conformations of the peptide shows that there is a rotation around the central part of the peptide at Lys^{345} upon peptide binding to rhodopsin. It is of interest that in a mutant G protein that is unable to couple to the β–adrenergic receptor contained in the S49 lymphoma cell line *unc* (for uncoupled from receptor), the homologous Arg^{390} is replaced by proline (Sullivan et al., 1987) that would not have such conformational flexibility. Thus this conformational change may be important in high-affinity binding of G proteins to activated receptors or in the receptor–catalyzed G protein activation process.

3. Implication for receptor–G protein interaction mechanism.

It was of interest to determine whether these studies on mechanisms of rhodopsin-G_t interaction would apply to mechanisms of interaction of other G proteins with their cognate receptors. Recent studies have shown that the carboxyl terminal undecapeptide of α_s, α_s-384-394, is able to selectively block β-adrenergic stimulation of adenylyl cyclase in saponin-permeabilized C6 glioma cells (Lazarevich et al., 1990; 1991). In addition, α_s-384-394 induces high–affinity β–agonist binding to β–adrenergic receptors, as determined by agonist/antagonist competition curves. The peptide has this effect even in the presence of guanine nucleotides, suggesting that it binds to and stabilizes the active receptor, analogous to the G_t peptide effect on rhodopsin.

Thus short peptide sequences carry a surprising degree of specificity of blocking (and mimicking) effects. Such studies suggest that these regions are important in the interactions of G_s with its cognate receptors. Current studies are directed at identifying commonalities of interaction sites and mechanism, expected on the basis of homologies between different receptors and G proteins, and detected in studies that show significant cross-talk between homologous receptors and G proteins. Of considerable interest also is the mechanism of specificity of receptor–G protein interaction. Synthetic peptide studies may be helpful in identifying unique sites of interaction that confer specificity of receptor–G protein interaction. Apropriate conformationally–constrained peptide mimetics of such regions may someday serve as therapeutic agents that selectively block particular receptor-G protein interactions.

References

Bennett, N. and Y. Dupont. (1985). The G-protein of retinal rod outer segments (transducin). Mechanism of interaction with rhodopsin and nucleotides. J. Biol. Chem. 260, 4156.

Campbell, A. P. and B. D. Sykes. (1989). Conformation of a Troponin I peptide bound to Troponin C as determined by ^{1}H NMR. In: *Calcium Protein Signaling* (H. Hidaka, ed.) Plenum Press.

Chabre, M. and P. Deterre. (1989). Molecular mechanism of visual transduction. Eur. J. Biochem. 179, 255–266.

Clore, G. M. and A. M. Groenenborn. (1982). Theory and applications of the transferred nuclear Overhauser effect to the study of the conformations of small ligand bound to proteins. J. Mag. Res. 48, 402–417.

Cote, T. E., E. A. Frey and R. D. Sekura. (1984). Altered activity of the inhibitory guanyl nucleotide-binding component (Ni) induced by pertussis toxin. J. Biol. Chem. 259, 8693–8698.

Dratz, E. A., J. F. Furstenau, H. E. Hamm and P. A. Hargrave. (1990). Probing the molecular mechanism of visual excitation using bioactive peptide sequences. Invest. Ophthalmol. Vis. Sci. 31/4, Suppl.79.

Dratz, E. A., H. E. Hamm, R. Zwolinski, J. Furstenau and C. Lambert. The 3D structure of the active interface between rhodopsin and G protein determined using 2D NMR. Biophys. J. 57, 1991.

Emeis, D., Kühn, H., Reichert, J. and Hofmann, K.P. (1982). Complex formation between metarhodopsin II and GTP-binding protein in bovine photoreceptor membranes leads to a shift of the photoproduct equilibrium. FEBS. Lett. 143, 29–34.

Gilman, A. G. G proteins: transducers of receptor–generated signals. Ann. Rev. Biochem. 56, 615–649, 1987.

Hamm, H. E., Deretic D., Arendt, A., Hargrave P. A., Koenig B., Hofmann K. P. (1988). Site of G protein binding to rhodopsin mapped with synthetic peptides from the α subunit. Science 241, 832–835.

Hamm, H. E., R. Zwolinski, H. M. Rarick, J. Furstenau and E. A. Dratz. Probing the structure of the rhodopsin binding domain of rod G protein. Biophys. J. 57, 1991.

Kurose, H., Katada, T., Amano, T. and M. Ui. (1983). Specific uncoupling by islet-activating protein, pertussis toxin, of negative signal transduction via α–adrenergic, cholinergic, and opiate receptors in neuroblastoma x glioma hybrid cells. J. Biol. Chem. 258, 4870.

Lazarevic, M., M. M. Rasenick and H. E. Hamm. Modulation of adenylyl cyclase activity by G protein specific synthetic peptides in C6 glioma cells. Soc. Neurosci. Abstr., 16, 1990.

Lazarevic, M., Watanabe, M., M. M. Rasenick and H. E. Hamm. Effects of site–specific peptides on blocking receptor–G protein interaction in permeable C6 glioma cells. Soc. Neurosci. Abstr., 17, 1991.

Liebman, P. A., Parker, K. R., and Dratz, E. A. 1987. The molecular mechanism of visual excitation and its relation to the structure and composition of the rod outer segment. Ann. Rev. Physiol. 49, 765–791.

Sullivan, K. A., Miller, R. T., Masters, S. B., Beiderman, B., Heideman, W. and H. R. Bourne. (1987). Identification of receptor contact site involved in receptor–G protein coupling. Nature (Lond.) 330, 758–762.

Vuong, T . M., M. Chabre and L. Stryer. (1984). Millisecond activation of trans-ducin in the cyclic nucleotide cascade of vision. Nature (Lond.) 311, 659–661.

Watanabe, M., Lazarevic, M., H. E. Hamm and M. M. Rasenick. Site–specific G_s peptides evoke high affinity binding to β–adrenergic receptors in permeable C6 glioma cells. Soc. Neurosci. Abstr., 17, 1991.

Wilden, U., S.W. Hall and H. Kühn. (1986). Phosphodiesterase activation by photoexcited rhodopsin is quenched when rhodopsin is phosphorylated and binds the intrinsic 48–kDa protein of rod outer segments. Proc. Natl. Acad. Sci. USA 83, 1174-11..

Wistow, G.J., Katial, A., Craft, C. and Shinohara, T. (1986). Sequence analysis of bovine retinal S-antigen. Relationships with α–transducin and G–proteins. FEBS Lett. 196, 23–27.

Wright, P. E., H. J. Dyson and R. A. Lerner. (1988). Conformation of peptide fragments of proteins in aqueous solution: Implications for initiation of protein folding. Biochem. 27, 1767–1775.

Comparative Studies on Chicken Skeletal and Smooth Muscle Dystrophins

Nathalie AUGIER°, Françoise PONS+, Jocelyne LEGER°, Roland HEILIG°°,

Agnès ROBERT°, Jean - J. LEGER °, and Dominique MORNET °

° Pathologie Moleculaire du muscle, INSERM U300, Faculté de Pharmacie, Av. Charles Flahault, 34070, Montpellier, Cedex,FRANCE.+ Pathologie Générale, Faculté de Médecine, Place Henri IV, 34000, Montpellier, FRANCE.°° Laboratoire de Génétique Moléculaire des Eucaryotes du CNRS, INSERM U184, 11 rue Humann, 67085, STRASBOURG, Cedex, FRANCE.

On the way to protein identification

The most common and severe human muscular dystrophy is the Duchenne type (DMD). DMD disease is caused by a defective gene (Murray et al., 1982; Davies et al., 1983) located on the short arm of the X chromosome (Xp 21). After Kunkel's group (Monaco et al., 1985) cloned the corresponding cDNA, major progress in identifying its protein product was achieved by the end of 1987. This new molecule was found to be of 3685 amino acid residues in humans (Koenig et al., 1988), but only 3660 amino acid residues in chicken skeletal muscles (Lemaire et al., 1988).

Abbreviation: DMD, Duchenne Muscular Dystrophy.

A dystrophin model has been derived from the primary structure of this molecule which contained four domains, and sequence homologies with other cytoskeletal proteins. Alpha actinin and spectrin share homologies with the N-terminal part of then dystrophin molecule, containing a putative actin binding site. The second domain is a rod-shaped central domain (Hammonds, 1987; Davidson and Critchley, 1988; Byers et al., 1989; Karinch, et al., 1990) containing four hinge segments which could give flexibility to the molecule (Mandel, 1989; Koenig and Kunkel, 1990), the third is a cystein rich domain with a putative calcium binding site. No strong predictable similarities were found within the C-terminal domain (Koenig et al., 1988). It is now possible to visualize dystrophin by immunofluorescence detection on cryostat sections of muscle fibers (Bonilla et al., 1988) and by the Western immunoblot procedure using specific antibodies against synthetic oligopeptides (Kao et al., 1988; Sugita et al., 1988) or recombinant proteins (Hoffman et al., 1987). This protein, which is lacking in DMD patients, is located within the muscle fibers membrane. Ultrastructural studies pinpointed dystrophin on the inner face of the sarcolemma (Zubrzycka-Gaarn et al., 1988; Watkins et al., 1988) with a repetitive periodicity of about 125 nm. But, dystrophin is not only restricted to muscular tissue, it is also expressed in brain tissue (Nudel et al., 1988; Nudel et al., 1989; Chelly et al., 1990).

Two attemps to isolate dystrophin have been made by other groups. Dystrophin was isolated in association with glycoproteins (Campbell et al., 1989; Ervasti et al. 1990) and in a denatured form (Matsumara et al., 1989). However, the precise function of this protein is still unknown. The objective here is to briefly describe our own attempts to isolate the dystrophin molecule and to understand its structural and functional properties. Electron microscopy and limited proteolysis approaches have allowed us to determine the general shape of the protein and to present evidence on tissue-specific isoforms of skeletal and smooth muscle dystrophins.

Dystrophin Isolation.

The first step towards dystrophin detection was to produce specific antibodies. Recombinant proteins expressed in E. coli corresponding to three different regions of the molecule (termed A-DYS, C-DYS and H-DYS for amino acid positions from 43 to 760, 1173 to 1728 and 3357-3660.) were used to produce polyclonal antibodies. Their affinities were tested by ELISA, with C-DYS having the highest affinity (Augier et al., 1990). All of these antibodies recognized both skeletal and smooth muscle dystrophin as a 400 kDa protein band on Western blot and were able to detect it at the periphery of muscular cells. We started isolating dystrophin from chicken gizzard smooth muscle because it was easier to extract this protein. Fresh muscles (10 to 100 mg) were rapidly mixed with a buffer solution (Tris HCl pH 9.0) in the presence of an inhibitor cocktail (0.5% Triton X100 and 10 mM iodoacetamide). The supernatant was adjusted to 20% ammonnium sulfate. The pellet was dissolved with a 0.1 M Tris buffer at pH 9.0 without Triton and clarified. A final purification was attained by affinity chromatography. The dystrophin solution was applied to an affinity column made with C-DYS antibody. The chromatography revealed that dystrophin was selectively retained, and eluted at acidic pH with a 43 kDa protein. The Western immunoblot pattern of this affinity column purification detected only one protein of about 400 kDa. As expected, and as found for skeletal muscle dystrophin (Knudson et al., 1988), the smooth muscle protein purified by this procedure corresponded to a protein band with a molecular mass of 400 kDa.

Dystrophin molecule shapes viewed by electron microscopy

Isolated smooth or skeletal muscle dystrophins were viewed by electron microscopy after rotatory shadowing the molecules. Different dystrophin

Gizzard   Skeletal

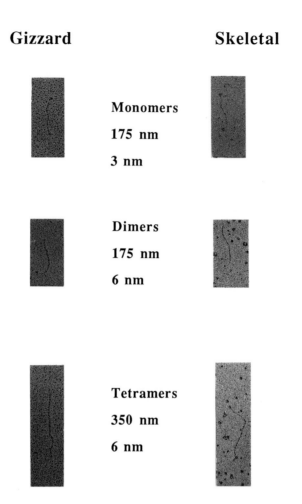

Monomers

175 nm

3 nm

Dimers

175 nm

6 nm

Tetramers

350 nm

6 nm

Figure 1: <u>Comparative electron microscopic images of smooth and skeletal muscle dystrophins.</u>

Dystrophin molecules have the same capacity of self-association. They have been found to be 175 nm long and 3nm thick monomers. When self-associating side by side, they were twice as thick as the monomer, and thus identified as dimers. Double-length structures were considered to be tetrameric forms. The general shapes of the smooth and skeletal muscle dystrophin images were quite similar.

shapes were obtained, thus indicating various structural arrangements of this molecule, similar to the structures described for alpha actinin and spectrin (Whiting et al., 1989). Dystrophin, like these other proteins is able to self-associate. The major species observed, which are shown in Fig.1, suggest that dystrophin monomers are elongated molecules of 175 nm (Pons et al., 1990), which can self-associate like side to side dimers. This structure can also be twice as long, thus representing an end to end dimer and tetramer formations . Larger so-called polymeric structures were also viewed. The same structure and arrangement was found in dystrophins isolated from smooth and skeletal muscles. Considering its structural properties and capacity of self-association, dystrophin could be compared to an elastic ribbon enveloping each muscular fiber.

Comparative proteolytic patterns of smooth and skeletal muscle dystrophin molecules

Limited proteolysis cleavage was used to investigate dystrophin substructure. This approach has also been used to detect the main structural domains of myosin or actin molecules (Offer et al., 1973; Mornet et al., 1984). Enzymes, which differed by their specificities, were used comparatively to split the smooth and skeletal dystrophin-enriched preparations. Time course studies of the proteolytic patterns were followed by electrophoretic analyses on gradient acrylamide slab gel (2.5 to 7%) in the presence of a denaturing agent and then transferred onto a nitrocellulose sheet . Protein bands were immunologically detected and located using A-DYS, C-DYS or H-DYS specific dystrophin antibodies. Proteolitic patterns obtained using various proteases differed from tissue to tissue. The major proteolytic products were identified and mapped as follows : limited proteolysis with trypsin gave rise to different C-terminal fragments of about 310 kDa for gizzard and 350 kDa for skeletal dystrophin, chymotrypsin produced N-terminal fragments of about 210 kDa for gizzard and 230 kDa for skeletal dystrophin, V8-protease gave shortened central fragments spanning 140 kDa for gizzard or 170 kDa for skeletal dystrophin. These results indicated the existence of two tissue-specific dystrophin isoforms. A limited proteolitic method should now be helpful for obtaining specific fragments whose functional properties could be investigated.

Molecular consequences of the structural and substructural arrangements of the dystrophin molecule.

Dystrophin is an elongated rod shaped molecule. It has a similar structure to that of alpha actinin and spectrin and the same capacity of self-association in oligomeric structures. Moreover, dystrophin molecules in association with other proteins and glycoproteins could be implicated in the complex meshwork anchored to the sarcolemmal membrane, and essential to muscle cell function.

Smooth and skeletal dystrophins are both isoforms. These isoforms could be the result of alternative splicing. Feener and others (1989) have described alternative splicing for the C-terminal part of the smooth muscle dystrophin. Our results suggest that alternative splicing is not restricted to the C-terminal part of the smooth muscle dystrophin but also involve the central part of the molecule.

Whether or not a dystrophin superfamily exists either as isoenzymes from a single gene or as dystrophin-like products, we assume that it is consistent with the presence of non homologous sequences within the N-terminal, C-terminal or central part of the protein. These molecules must differ in function. More specifically, they must differ in their affinities to actin , in their capacities of self association and/ or in their ability to fix to the membrane via different proteins. These hypotheses would indicate a complex structural role played by the dystrophin molecule in normal muscles and also in non-muscle tissues such as brain tissues (Clerk et al., 1990).

Acknoledgements: This work was supported by INSERM, CNRS and the Association Française contre les Myopathies, (A.M.F.).N. Augier is awarded with an AMF fellowship.

References.

- AUGIER, N., HEILIG, R., BOUCRAUT, J., LEGER, Joc., LEMAIRE, C., GEORGESCO, M., ESCHENNE, B., MANDEL, J.L., PONS, F., PELLISSIER, J.F. and LEGER, J. (1990), J. of Clin. Inves., in press.
- BONILLA, E., SAMITT, C.E., MIRANDA, A.F., MAYS, A.P., SALVIATI, G., DIMAURO, S., KUNKEL, L.M., HOFFMAN, E.P. and ROWLAND, C.P., (1988), Cell, **54**, 447 - 452.
- BYERS, T.J., HUSAINCHIST, A., DUBREUIL, R.R., BRANTON, D. and GOLDSTEIN, L.S.B. (1989), J.Cell. Biol., **109**, 1633 - 1641.
- CAMPBELL, K.P. and KAHL, S.D., (1989), Nature, **338**, 259 - 262.
- CHELLY, J., HAMARD, G., KOULAKOFF, A., KAPLAN, J.C., KAHN, A. and BERWALD-NETTER, J. (1990), Nature, **344**, 64 - 65.
- CLERK, A., MUNTONI, F. and STRONG, P., (1990), Biochem. Soc. Trans., **18**, 388 - 389.
- DAVIDSON, M.D. and CRITCHLEY, D.R. (1988), Cell **52**, 159 - 160.
- DAVIES, K.E., PEARSON, P.L., HARPER, P.S., MURRAY, J.M., 0'BRIEN, T., SARFARAZI, M. and WILLIAMSON, R. (1983), Nucleic Acids Res. **11**, 2303 - 2312.

- ERVASTI, J.M., OHLENDIECK, K., KAHL,S.D., GAVER, M.G. and CAMPBELL, K.L.P. (1990), Nature, **345**, 315 - 319

- FEENER, C., KOENIG, M. and KUNKEL, L.M. (1989), Nature **338**, 509 - 511.

- HAMMONDS, R.G. Jr. (1987), Cell, **51**, 1.

- HOFFMAN, E.P., BROWN, R.H. and KUNKEL, L.M. (1987), Cell, **51**, 919 - 928.

- KAO, L., DRSTENANSKY, J., MENDEL, L.J., RAMMOHAN, K.W. and GRUENSTEIN, E. (1988), Proc. Natl. Acad. Sci. USA , **85**, 4491- 4495.

- KARINCH, A.M., ZIMMER, W.E., GOODMAN, S.R. (1990), J. Biol. Chem., **265**, 11833 - 11840.

- KNUDSON, C.M., HOFFMAN, E.P., KAHL, S.D., KUNKEL, L.M. and CAMPBELL, K.P. (1988), J. Biol. Chem., **263**, 8480 - 8484.

- KOENIG, M., MONACO, A.P. and KUNKEL, L.M. (1988), Cell, **53**, 219 - 228.

- KOENIG, M., and KUNKEL, L.M. (1990) , J. Biol. Chem., **265**, 4560 - 4566.

- LEMAIRE, C., HEILIG, R. and MANDEL, J.L. (1988), EMBO J., **7**, 4157 - 4162.

- MANDEL, J.L. (1989), Nature , **339**, 584 - 586.

- MATSUMARA,K., SHIMIZU, T. and MANNEN, T., (1989), Biomed. Res., **10**, 325 - 328.

- MONACO, A.P., BERTELSON, C.J., MIDDLESWORTH, W., COLLETTI, C.A., ALDRIDGE, J., FISCHBECK, K.H., BARLETT, R., PERICAK-VANCE, Roses, A.D. and KUNKEL, L.M. (1985) , Nature ,**316**, 842 - 845.

- MORNET, D., UE, K. and MORALES, M.F. (1984), Proc. Natl. Acad. Sci. U.S.A., **81**, 736 - 739

- MURRAY, J.M., DAVIES, K.E., HARPER, P.S., MEREDITH, L., MUELLER, C.R. and WILLIAMSON, R., (1982), Nature, **300**, 69 - 71.

- NUDEL, U., ZUK, D., ZEELON, E., LEVY, Z., NEUMAN, S. and YAFFE, D. (1989), Nature , **337**, 76 - 78.

- NUDEL, U., ROBZIK, K. and YAFFE, D. (1988), Nature, **331**, 635 - 638.

- OFFER, G., MOOS, C. and STARR, R. (1973), J. Mol. Biol., **74**, 653 - 676.

- PONS, F., AUGIER, N., HEILIG, R., LEGER, Joc., MORNET, D. and LEGER, J.J., (1990), Proc. Natl. Acad. Sci. U.S.A. , **87**, 7851-7855.

- SUGITA, H., ARAHATA, K., ISHIGURO, T., SUHARA, Y., ISHIURA, S., EGUCHI, C., NONAKA, I. and ZAWA, E. (1988), Proc. Jpn. Acad., **64**, 37-39.

- WATKINS, S.C., HOFFMAN, E.P., SLAYTER, A.S. and KUNKEL, L.M. (1988), Nature, **333**, 863-866.

- WHITING, A., WARDALE, J. and TRINICK, J., (1989), J. Mol. Biol., **205**, 263-268.

- ZUBRZYCKA-GAARN, E. E., BULMAN, D.E., KARPATI, G., BURGHES, A.H.M., BELFALL, B., KLAMUT, H.J., TALBOT, J., HODGES, R.S., RAY, P.N. and WORTON, R.G. (1988), Nature, **333**, 466-469.

Use of Synthetic Peptides in the Study of the Function of Dystrophin

S.V. Perry, B.A. Levine*, A.J.G. Moir**, V.B. Patchell
Department of Physiology
University of Birmingham
Birmingham B15 2TT
UK

From the work of the Kunkel (Koenig et al., 1987) and Worton groups (Burgess et al., 1987) it has been established that the protein dystrophin is absent or in some cases present in a modified form in patients with Duchenne and Becker muscular dystrophies (Xp21 myopathies). The amino acid sequence deduced from analysis of the complementary DNA indicates that dystrophin from human skeletal muscle has a molecular weight of 427,000 and consists of a polypeptide chain of 3,865 residues with a multi-domain structure (Koenig et al., 1988). The N-terminal domain of 240 residues shows homology to the N-terminal domain of the α-actinins in which region actin has been shown to bind (Mimura & Asano, 1986). The second and largest domain is formed by a succession of 25 triple helical segments (the fourth of which is truncated) with loose homology between them and which show weak similarity to similar repeats in α and β-spectrin. A third domain extending over 150 residues is cysteine-rich and is similar in sequence to the carboxy terminal domain of α-actinin that contains two calcium binding sites. The corresponding sites on dystrophin are rather rudimentary and may not be functional. The carboxy terminal domain consists of 420 residues and does not possess homology to any currently known protein.

* Department of Inorganic Chemistry, University of Oxford, Oxford; ** Department of Molecular Biology and Biotechnology, University of Sheffield, Sheffield, UK.

The conclusion from study of the cDNA sequence that the molecule is rodlike has been confirmed by electron microscope studies of the isolated protein from which values for the molecular length of 180 nm (Pons et al., 1990) and 100–120 nm (Shimizu et al., 1990) have been reported. These values should be compared with the periodicity of 120 nm of antibody staining reported from immunocytochemical studies (Cullen et al., 1990,a,b) which probably corresponds to the apparent length of the molecule associated with the membrane. The latter authors conclude that dystrophin forms a network structure with the rod-like region lying 15 nm from the cytoplasmic face of the membrane and parallel to it with the C-terminal region actually within the membrane. These observations, the long recognised increase in muscle permeability that occurs in Duchenne muscular dystrophy and the recently reported changes in ion permeability that occur on stretching muscle from the mdx mouse, but not in the normal control (Franco & Lansman, 1990), strongly indicate a function for dystrophin in muscle membrane structure.

In view of its homology with the cytoskeletal protein spectrin it is likely that dystrophin has, in part at least, a similar role to the former protein. Much of what we know of the role of spectrin is derived from study of the erythrocyte cell membrane with which is associated an elaborate cytoskeletal system. This system has evolved to be flexible and resilient so that the erythrocyte survives the rigors of the circulation. Similarly, but unlike most other cells, the muscle fibre membrane must be also strengthened to maintain its unique properties despite the stresses imposed upon it as a result of contractile activity. Our approach therefore is to use the erythrocyte system as a paradigm to determine which of the known membrane integral and cytoskeletal proteins interact with dystrophin and identify the sites involved in those cases in which specific interactions can be demonstrated. In Table 1 the major protein components of the erythrocyte membrane and cytoskeleton, most of which have at some time been reported to interact with spectrin, are listed.

From the homology of the N-terminal sequence of dystrophin with α-actinin, it had been assumed, but not directly proven, that actin interacts with dystrophin. With the aim of

demonstrating such an interaction and defining the sites
involved we employed high resolution proton NMR spectroscopy in

Protein	Subunit M_r	Structure	Copies/cell
Integral components			
Band 3	90000	dimer/tetramer	1000000
Glycophorins	23000-31000	dimer	500000
Skeleton components			
Spectrin	240000	tetramer	100000
	220000		
Band 4.1	78000	monomer	100000
Band 4.9	45000	trimer	50000
Actin	43000	oligomer	500000
Tropomyosin	29000	dimer	70000
	27000		

Table 1 Protein components of the erythrocyte membrane and
membrane cytoskeleton. After Carraway & Carraway (1989)

the manner similar to that used to determine the sites involved
in the interactions between actin and the regulatory proteins,
troponin I (Levine et al., 1988) and caldesmon (Levine et al.,
1990b). In the latter studies the changes in the proton
resonance signals of synthetic peptides corresponding to the
binding sites on the intact protein that occurred on
interaction were used to identify the amino acid side chains
involved. Likewise our approach has been to study the
interaction of F-actin with synthetic peptides corresponding to
regions of the dystrophin molecule (Levine et al., 1990a).
Unlike the studies with the regulatory proteins where there is
direct biochemical evidence defining the location of the
actin-binding sites, in the case of dystrophin the peptides
synthesised corresponded to putative actin-binding sites and
regions of special interest to us.

A series of synthetic peptides corresponding to residues
1-32, 10-32, 128-156, 154-182 and 3429-3440 of dystrophin were
prepared, purified and dissolved in D_2O containing 0.5 mM
dithiothreitol and 10 mM sodium bicarbonate, pH 7.0. To
solutions of the peptides (150-200 µM) were titrated samples of
a stock solution of F-actin (200 µM) to bring the final actin
concentration to 16 µM. Proton NMR spectroscopy was carried

out on a Bruker 500 MHZ instrument.

The N-terminal peptide corresponding to residues 1-32 was rather insoluble but 160 µM solutions gave satisfactory signals. The intensities however were broad, particularly those due to hydrophobic residues and did not correspond well to the composition since the majority did not integrate properly indicating multimeric aggregates (Fig 1). On addition

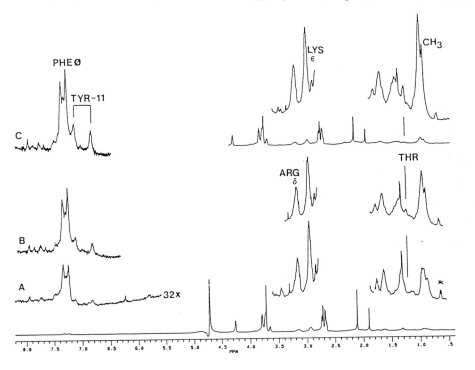

<u>Figure 1</u> ^{1}H NMR spectral effects of the binding of F-actin to dystrophin residues 1-32. (A) Dystrophin peptide, 160 µM, in D_2O solution. Note the very broad lineshapes of the peptide resonances shown on an expanded scale (32x). (B) Addition of 8 µM F-actin causes an apparent increase in signal intensity (cf the signal of the standard denoted by a *) resulting from a decrease in resonance linewidth. (C) In the presence of 16 µM F-actin, the resonances of, eg Tyr-11 and Thr-20,22 are now readily resolved (From Levine et al., 1990a).

of F-actin up to 16 µM, at which concentration actin does not contribute significantly to the spectrum (see Fig 2D), marked changes in signal intensity occurred progressively as the actin concentration rose. The intensities of the signals arising

from a number of hydrophobic side chains increased implying
that disaggregation of the peptide was occurring. At the same
time some formation of precipitate was observed. The
observations were compatible with disaggregation of the peptide
occurring with interaction with actin but interpretation of the
spectral effects was complicated by the fact that the two
changes were taking place simultaneously.

The observation that the signals from tyrosine, which
residue is located in the first eleven residues of the peptide,
were broadened in the absence of F-actin whereas those due to
phenylalanine (residues 21, 29 and 32) were not, suggested that
aggregation of peptide 1-32 involved the N-terminal region.
This conclusion was confirmed by examination of the proton NMR
spectrum of the peptide consisting of residues 10-32. Well
resolved sharp signals were obtained for all residues
indicating that the peptide was not aggregated. Examination of

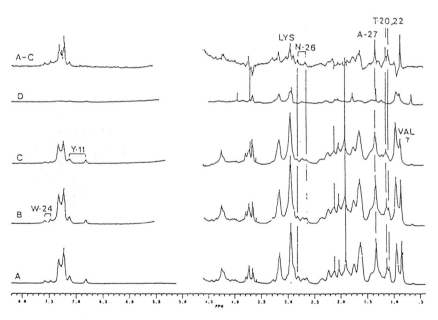

Figure 2 ^{1}H NMR spectral titration of dystrophin residues
10-32 with F-actin. (A) 200 μM dystrophin peptide. (B) and
(C) upon addition of 8 μM and 16 μM F-actin respectively (cf
Figure 1). (D) spectrum of the stock solution of F-actin
titrated into the peptide. (A-C) difference spectrum resolving
signals of the peptide perturbed by complex formation with
F-actin. The resonance assignments shown were obtained by
analysis of the COSY spectrum. (From Levine et al., 1990a)

the signals by two dimensional correlated spectroscopy enabled
the assignments shown in Fig 2 to be derived. Addition of
F-actin to the peptide resulted in selective broadening of the
signals arising from phenylalanine 21, β CH$_3$ of alanine 27,
γ CH$_3$ of threonines 20 and 22, γ CH$_3$ of valine 25 and of ε CH$_2$
of lysine. It was therefore concluded that the reduction in
the segmental mobilities of the side chains of residues 18-27
and possibly those of additional C-terminal residues was due to
interaction with F-actin at this site.

Similar experiments to those illustrated in Fig 2, but in
which peptide 10-32 was replaced by peptide 3429-3440, were
carried out. In this case there is no evidence of interaction
with F-actin. The possibility that peptide 10-32 had a general
property of binding to proteins was excluded by the failure to
demonstrate any interaction with serum albumin. These control
experiments indicated the specific nature of the binding site
on peptide 10-32. When the amino acid sequences of the
putative actin-binding regions of four different α-actinins
(Blanchard et al., 1989), β-spectrin regions (Karinch et al.,
1990) and gelation factor (Noegel et al., 1989) are aligned
alongside the actin-binding site of dystrophin the sequence
lys-thr-phe-thr, which corresponds to residues 19 to 22 in
human dystrophin, is also present in a homologous position in
each of the six other proteins (Fig 3). There is also
considerable homology with several residues on either side of
this tetrapeptide sequence in all the proteins. We conclude
that as is the case with dystrophin the tetrapeptide forms part
of the actin-binding site of the α-actinins, β-spectrin and
gelation factor. It is of interest that the spectrin-like
domain of human dystrophin does not contain the tetrapeptide
sequence. Thus in dystrophin the actin-binding site is not
located in the spectrin-like domain but at the extended
N-terminus of the molecule.

Much has been learnt about the amino acid residues
involved in the docking sites on the actin molecule for myosin
and the thin filament proteins concerned in the regulation of
contraction. The recent determination of the tertiary
structure of actin (Kabsch et al., 1990) has enabled the
precise location of these sites on the molecule to be made. In
its role as a cytoskeletal protein actin forms complexes with

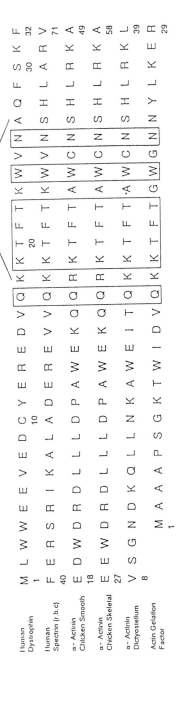

Figure 3 Comparison of amino acid sequences of the actin-binding domains of human dystrophin, human β-spectrin, chicken α-actinins and actin gelation factor from dictyostelium

many other proteins and it is interesting to compare the structure of these complexes with those involved in the contraction-relaxation cycle. We have demonstrated that troponin I interacts with residues 1-7 and 23-26 (Levine et al., 1988) and that caldesmon, which binds to the same N-terminal region as troponin I, also binds to another site which is slightly further along the polypeptide chain than the second troponin I-binding site, in the region where S1-binding has been demonstrated (Levine et al, 1990b). Thus in effect the N-terminal 40 residues of actin have a functional role in the contraction relaxation cycle. To define the dystrophin-binding site on actin the NMR studies require that actin peptides are titrated with dystrophin or substantial fragments of it that will form complexes with binding constants of $>10^4$. As a preliminary approach until dystrophin or soluble fragments of the molecule become available in the amounts required for this type of experiment, dystrophin peptide consisting of residues 10-32 was titrated against peptides of actin representing putative binding sites. Caution must be exercised in drawing firm conclusions from these experiments as the binding constants of any complexes formed are likely to be low and not necessarily detected under conditions of the NMR experiments. Nevertheless the results suggest that dystrophin interacts with the smaller domain of actin but probably not at the sites at which caldesmon and troponin I bind, i.e. the region represented by the N-terminal 40 residues. In confirmation of the latter conclusion it was shown by proton NMR spectroscopy that actin interacts with the dystrophin peptide 10-32 in the presence of the inhibitory peptide isolated from troponin I (residues 97-119) which is known to interact with actin at two N-terminal sites. Further evidence that troponin I and dystrophin bind to different sites on actin was given by the observation that dystrophin peptide 10-32 had no effect on the inhibition of the Mg^{2+}-activated ATPase of actomyosin by troponin I.

Although dystrophin does not replace spectrin in muscle the homology of a large part of the molecule with spectrin, coupled with its ability to bind to actin, suggests a similar function. Its role in the cell will be modulated by the presence of the C-terminal domain which appears to be unique to

this protein and the part of the molecule most strongly
conserved in evolution (Lemaire et al., 1988). If dystrophin
exists as an anti-parallel dimer (or tetramer) then the
actin-binding sites would resemble those of spectrin which is
usually represented in cytoskeletal schemes as binding actin at
either end of the molecule (Carraway & Carraway, 1989). The
NMR studies indicate that the N-terminal 11 amino acids are
involved in the aggregation behaviour of the dystrophin peptide
1-32, leaving the C-terminal residues with segmental mobility.
If this N-terminal region is important in determining the
orientation for dimer formation, the molecules could take up a
parallel side-by-side arrangement with N and C-termini at
different ends of the molecule or aggregate as an elongated
dimer with the N-termini in the centre and the C-termini at
each end. It should be stressed however that the peptide
studied represents less than 1% of the whole molecule and
polymer formation will no doubt depend on interactions between
other parts of the molecule with each other and the N-terminal
residues. More detailed information as to how the dystrophin
molecule associates with itself and other components of the
cytoskeleton and membrane must await the progress of other
experiments now in progress.

REFERENCES

Blanchard A, Ohanian V, Critchley D (1989) The structure and
    function of α-actinin. J Musc Res Cell Motility 10:280-289
Burgess AHM, Logan C, Hu X, Befall, B, Worton C, Ray PN (1987)
    A cDNA clone from the Duchenne/Becker muscular dystrophy
    gene. Nature Lond. 328:434-437.
Carraway KL, Carraway CAC (1989) Membrane cytoskeletal
    interactions in animal cells. Biochim Biophys Acta
    988:147-171
Cullen MJ, Walsh J, Nicholson LVB, Harris JB (1990a)
    Ultrastructural localization of dystrophin in human muscle
    by using gold immunolabelling. Proc Roy Soc Lond B
    240:197-210
Cullen MJ, Walsh JV, Nicholson LVB, Zubrzycka-Gaarn EE, Ray PN,
    Worton RG, Harris JB (1990b) Ultrastructural localization
    of dystrophin in human muscle using antibodies to
    different segments of the dystrophin molecule. J Neur Sci
    98(Supplement):255
Franco A, Lansman JB (1990) Calcium entry through stretch
    inactivated ion channels in mdx myotubes. Nature
    344:670-673
Kabsch W, Mannherz HG, Suck D, Pai EF, Holmes KC (1990) Atomic
    structure of actin: DNase 1 complex. Nature 347:37-44.

Karinch AM, Zimmer WE, Goodman SR (1990) The identification and sequence of the actin-binding domain of human red blood cell β-spectrin. J Biol Chem 265:11833-11840

Lemaire C, Heilig R, Mandel J L (1988) The chicken dystrophin cDNA: sticking conservation of the C-terminal coding and 3' untranslated regions between man and chicken, Embo J 7:4157-4162

Levine BA, Moir AJG, Perry SV (1988) The interaction of troponin I with the N-terminal region of actin. Eur J Biochem 153:389-397

Levine BA, Moir AJG, Patchell VB, Perry SV (1990a) The interaction of actin with dystrophin. FEBS Letters 263:159-162

Levine BA, Moir AJG, Audemard E, Mornet D, Patchell VB, Perry SV (1990b) Structural study of gizzard caldesmon and its interaction with actin. Eur J Biochem 193:687-696

Koenig M, Monaco AP, Kunkel LM (1988) The complete sequence of dystrophin predicts a rod-shaped molecule cytoskeletal protein. Cell 53:219-228

Koenig M, Hoffman EP, Bertelson CJ, Monaco AP, Feener C, Kunkel LM (1987) Complete cloning of Duchenne muscular dystrophy cDNA and the preliminary genomic organisation of the DMD gene on normal and affected individuals. Cell 50:509-517

Mimura N, Asano A (1986) Isolation and characterization of a conserved actin-binding domain from rat hepatic actinogelin, rat skeletal muscle and chicken gizzard α-actinins. J Biol Chem 261:10680-10687

Noegel AA, Rapp S, Lottspeich F, Schleicher M, Stewart M (1989). The Dictyostelium gelation factor share a putative actin-binding site with α-actinins and dystrophin and also had a rod domain containing six 100 residue motifs that appear have a cross-beta conformation. J Cell Biol 109:607-618

Pons F, Augier N, Heilig R, Leger J, Mornet D, Leger JJ (1990) Isolated dystrophin molecules as seen by electron microscopy. Proc. Natl. Acad. Sc. 87:7851-7855

Shimizu T, Murayama T, Sunada Y, Mannen T, Ozawa E, Maruyama K (1990) Immunochemical localization and electron microscope visualization of dystrophin. J Neur Sci 98(Supplement):230

Subject Index